Reactive Power Compensation

By

Dr Hidaia Mahmood Alassouli ·

Reactive Power Compensation

The project examined and corrected by

Dr Hidaia Mahmood Alassouli

Project done by the Student Muzi Dlamini

Electrical and Electronic Engineering

Department University of Swaziland

ABSTRACT

Power systems have two components of apparent power: active and reactive power. Both components are necessary for functioning of electrical systems. The active power is the average power absorbed by the resistive load. The reactive power is the measure of energy exchange between the source and reactive power of load. Energy storage devices do not dissipate or supply power, but exchange power with the rest of system.

Active power is the one that is converted to other forms of energy in the load yet reactive power is only responsible for magnetizing purposes. Power factor is a ratio depicting how much of the power supplied is real. The reactive current contribute in the value of the overall magnitude of current in transmission lines causing unnecessarily high line currents and low power factor.

Since a low power factor means higher amount of apparent power need to be supplied by the utility company, thus the company must also use bigger generators, large transformers and thicker transmission/distribution lines. This requires a higher capital expenditure and operational cost which usually result in the cost being passed to the consumer.

In this research, we seek to identify the effects of a low power factor on Swaziland Electricity Company's power supply system and recommend possible solutions to the problem. The results are useful in determining how to optimally deliver power to a load at a power factor that is reasonably close to unity, thus reducing the utility's operational costs while increasing the quality of the service being supplied.

ACKNOWLEDGEMENTS

This project would not have been a success without the mutual help and guidance of my academic and industrial supervisors, Mr. J.S. Manong'a and Mr. E.S. Mkhonta.

I would also like to appreciate:

The Swaziland Electricity Company for providing me with all the resources and training which were required during the project.

My family for their support and motivation.

My colleagues who gave suggestions for the improvement of some parts of the project.

Mr. M. Maziya and Mrs. H. Hlophe who assisted on simulations and all the SEC project staff.

The almighty God for always keeping me strong and safe while working on this project.

TABLE OF CONTENTS

CHAPTER 1: INTRODUCTION

1.1 Introduction to Research

AC electrical power system loads have resistive and reactive impedances. The electricity supply network therefore possesses an active and a reactive power component as a result of the characteristic of the load impendences [1]. The reactive component can be further subdivided into two states: a leading and lagging state. Leading reactive power comes as a result of the capacitive component of the load whereas lagging reactive power comes due to the inductive component of the load introducing a lagging phase shift in the network.

This gets us to the topic of interest, 'Power Factor'. It is a ratio that tells us how much power from that being supplied by the utility is being actually used to do useful work by the customer. Figure 1.0 below shows this relationship in a form of a right angled power triangle.

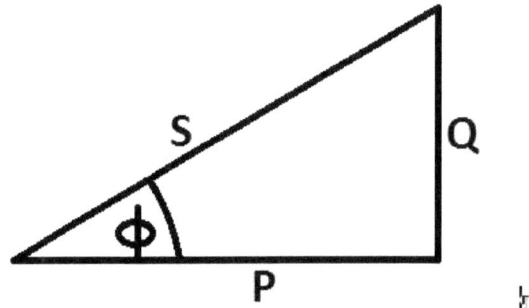

Figure 1.0: The Power Triangle [2].

Where:

S = apparent power (VA)

Q = reactive power (VAR)

P = real power (W)

Apparent power (S) is a complex combination of real power (P) and reactive power (Q). Real power, also called productive power does useful work and the reactive component of the power generates magnetic fields necessary for the operation of inductive devices such as AC motors, transformers etc. When in excess, reactive power can become detrimental to a power system as it greatly reduces the power factor, thus decreasing the distribution capacity while increasing the operational costs of the utility company.

1.2 Objectives

The main objectives of this project are:

- Identify the causes of a low PF in SEC's power system paying special attention to the eastern side of the network.

- Calculate the power factor and the component values for the affected substations of SEC's network.

- Evaluate the effects of a low power factor in SEC's network.

- Draw conclusions and make recommendations for any modifications or improvements on PFC.

1.3 Significance of Research

Whenever a power system supplies power at a low power factor, a significant percentage of the power doesn't do useful work at the load. This implies that the network is carrying higher current than required by load to cater for amount of reactive power in the network. The high transmission/distribution current result in increased system losses and voltage drops, thus reducing the power system's capacity. Since Swaziland is currently importing around 70% of its power from neighboring countries in order to meet the demand of its customers, so correcting power factor will help in reducing this deficit. This research will serve as a guide on how to keep a power network operating at a power factor close to unity. Though this research is based mainly on SEC's power system as it is where most of the work was done, the results are actually valid for nearly all power systems.

1.4 Problem Statement

The network of SEC has two injection points, Mahamba and Edwaleni. Mahamba supports the Eastern network while Edwaleni II is in the Western network. The customers in the Eastern network are mainly the agriculture industry with their major load being pumps for irrigation. In the west customers are nearly balanced between industrial and domestic, the load is mainly heating equipment, small motors, ventilation and cooling systems. This explains why the power factor of SEC's network is not optimal in the eastern side of the system as it is extremely loaded with reactive power. With the company importing around 70% of its power from its neighboring countries, power factor problems should be always properly dealt with to ensure that what we have is utilized efficiently while trying to breach the gap on the amount that is being bought from Mozambique and South Africa.

Moreover, not only the system losses are increased when the system is operating at a low power factor, the voltage stability is also disturbed causing a drop in quality of supply as some electronic equipment are sensitive to voltage changes. Typically, the costs involved in installing power factor correction are paid back within a year or two and after that the savings will actually reduce the operational costs. On the other hand over-correction should be always avoided. This can be done through a thorough study and collection of enough relevant data about the network's power factor before finalizing on a suitable method of correction and deciding on the component values.

1.5 Hypothesis

If we know the factors that contribute to the power factor being low and the load behavior of the affected portion of the network, then we can choose the appropriate PFC method to minimize the effects of those loads.

1.6 Layout of Thesis

Chapter two of this thesis discusses literature and theories behind power factor and its correction. Here we include the causes and effects of a low power factor in detail and how it is usually corrected in a power system.

In chapter three the research methodology used in the study will be described. The setting and the study design are also described. Here also the methods used to collect the data and the way the data will be analyzed are explained.

In chapter four, data on loading of five of SEC's substations will be collected and used in calculating their power factor and the approximated size of correction equipment will be determined from the results.

Chapter five consist of the simulation model where the load behavior will be simulated and also the simulation results will be discussed here. The sixth chapter reviews the benefits of PFC and evaluates the cost and benefits associated with this project. Finally, chapter seven summarizes the research and also reveals the recommendations and conclusions made.

CHAPTER 2: LITERATURE REVIEW

2.1 Introduction

This chapter discusses the causes and effects of a low value of PF in an AC power network's performance. It also outlines the need for power factor correction in affected parts of the power system and reviews the principles of operation for each of the methods of PFC discussed by the literature.

2.2 Power Factor

Power factor is the cosine of the angle between voltage and current of the power system. This angle is also equivalent to the angle of the network's impedance. Consider the power triangle (Fig. 1.0), the power factor can be also expressed as a ratio of real power to apparent power. i.e.

$$PF = \frac{Real\ Power}{Apparent\ Power} \quad \text{---} (2.0)$$

$$= \cos \Phi$$

Where

- **Apparent power** (S) is the complex summation of real and reactive power. It is a useful means of rating power equipment [3]. It can be further expressed as also a product of the circuit's current and voltage or the square of current multiplied by the circuit's impedance. Apparent power is measured in volt amperes (VA) as shown by the equations below.

$$S = P + jQ \quad \text{---} (2.1)$$

$$= I^2Z \quad \text{---} (2.2)$$

$$= VI^* \quad\text{---} (2.3)$$

- **Real power** (P), sometimes called active power, is the power dissipated by the load measured in watts (W). It is expressed as a product of the square of current and the circuit's resistive impendence or power factor multiplied by apparent power as shown in the equations below.

$$P = I^2R \quad\text{---} (2.4)$$
$$= S\,(\cos\Phi) \quad\text{---} (2.5)$$

- **Reactive power** (Q) is the component that is responsible for generation of magnetic fields for production of flux which is important for the operation of inductive devices. Mathematically, it is given by the product of the square of current and the reactive impendence or the apparent power multiplied by the sine of the power angle. This is shown in the equations below.

$$Q = I^2X \quad\text{---} (2.6)$$
$$= S\,(\sin\Phi) \quad\text{---} (2.7)$$

Real power is always positive while reactive power is negative for capacitive reactance and positive for inductive reactance [4]. This is because capacitive reactive power is injected to the network yet inductive reactive power is absorbed by the load. The power factor is a unit-less quantity and its value can either be expressed as a ratio or a percentage. It varies between zero and one i.e.

$$0 \leq PF \leq 1$$

The value of the PF is only unity when both the current and voltage are in phase i.e. purely resistive circuit. If different from unity, it can either be leading (for a capacitive load) or lagging (for an inductive circuit).

2.3 Causes of a low PF

A low PF in an AC network generally results from an excessive amount of reactive power drawn by inductive loads and some distribution equipment such as transformers. Transmission lines which are used in transporting electric power from generation units through the distribution system to the load exhibit electrical properties. These include resistance, inductance, capacitance and conductance. The lengthy nature of transmission lines gives them high line inductance. SIL can also be used to explain the causes of low power factor.

$$SIL = \frac{3|V_R|^2}{Z_C} \quad \text{- -} \quad (2.8)$$

Where V_R is the receiving end voltage and Z_C is the characteristic impedance (surge impedance) of the line. In lossless lines resistance and conductance are neglected i.e.

$$Zc = \sqrt{\frac{jwL}{jwC}} = \sqrt{\frac{L}{C}} \quad \text{- -} \quad (2.9)$$

With L and C being the inductance and capacitance of the transmission line.

If transmission line is loaded to its surge impedance, the reactive power in the sending end and receiving end equals zero (unity power factor). However, below SIL the line possesses a capacitive power factor while at heavy loading (above SIL) it becomes inductive.

Inductive loads require additional current for the creation of a magnetic field. Some of the equipment that lower PF are listed in Table 2.0.

Typical unimproved power factor by equipment

Equipment	Power Factor (%)
Air Compressors	75 – 80
Hermetic Motors	50 – 80
Arc Welding	35 – 60

Resistance Welding	40 – 60
Machining	40 – 65
Arc Furnaces	75 – 90
Standard Stamping	60 – 70
High speed Stamping	45 – 60
Spraying	60 – 65

Table 2.0: Power factor by selected equipment [5].

Most equipment tend to operate at a low PF during minimum load while the highest PF occurs at full load. Phase-angle-controlled loads like VSDs can also result in the absorption of reactive power.

2.4 Disadvantages of a low power factor

A low PF results in increased current in power lines which then introduces many drawbacks to a power system. Some of them are explained below:

1. **Loss of efficiency of equipment**

Whenever a distribution equipment is operating at a significantly low PF, the amount of useful power (active power) it can carry is reduced due to amount of reactive power contained in the device. Equipment efficiency is also compromised by the presence of high line currents. Equation 2.10 shows the inverse relation between PF and line current.

$$I_L = P/(\sqrt{3}V_L * PF) \text{ -------------------------------------- (2.10)}$$

Where: I_L, P, and V_L are line current, active power and line to line voltage, respectively.

Example 1: for a 33MW system operating at 66kV line to line

- At PF= 0.5, the line current is 577.35A
- At PF=1.00, the line current is 288.67A

Note the significant drop in current when power factor is raised from 0.5 to 1.00.

2. **Increased system losses**

The chief source of losses in any transmission or distribution line are the heat losses in the conductors when the current tries to overcome the ohmic resistance. It is also known as 'I squared R', indicating that it is proportional to the square of current being transmitted in the power line. A low PF is usually associated with high inductive loads. These cause an increase power losses by increasing the reactive component of load current meaning that more current will need to be supplied to meet a constant real power demand.

The power losses equation is given below.

$$P_{losses} = I_L^2 R \ \text{---} (2.11)$$

Where I_L and R are the current and resistance of the line.

Example 2: Extending the previous example for line resistance of 10Ω we get.

At PF=0.50, power losses are 333KW.

At PF=1.00, power losses are 83KW.

Also note the inverse relation between PF and power losses (I^2R).

3. Unnecessarily large investments

A power system operating at a low PF requires that larger equipment i.e. highly rated generators, transformers, and conductors to meet an arguably lower real power demand. For instance if a load requires 1000KW, it will need a 1000KVA transformer at unity PF, but that won't be the case for when it is different from unity. See table 2.1.

Required transformer rating for 1MW load at different power factors

Power Factor	Transformer Rating (KVA)
1	1000
0.8	1250
0.6	1667
0.4	2500
0.2	5000

Table 2.1: Effect of power factor on required transformer rating.

Despite the fact that reactive power is vital in the operation of some equipment, table 2.1 shows how a very low system power factor can negatively affect the sizing of power utility transformers. Also since a low PF is associated with high line currents meaning transmission lines will require thicker cables. Thus a low PF increases the capital cost for the power utility by increasing the required rating of the systems equipment for relatively lower load demands.

4. Excessive voltage drops in transmission lines

Large currents flowing through transmission conductors during low PF conditions cause high voltage drops across the transmission line which is proportional to the impedance of the line. Extreme voltage drops in transmission lines trigger under-voltage relays to induce over-tapping in distribution transformers as they attempt to raise the voltage to its standard levels. Most electrical equipment is designed to operate in a certain fixed range of voltages in such a way that if power is supplied at a lower voltage, the device can either be damaged or operate incorrectly.

Voltage drops in a transmission line are given by

$$V_{drop} = I_L R \qquad\qquad\qquad\qquad\qquad\qquad\qquad\qquad (2.12)$$

Where I_L and R are the current and resistance of the line.

2.5 Power Factor Correction Techniques

The problems associated with a low value of PF brings about the need for correction. This can be easily done by adding capacitive reactance to correct a lagging PF and inductive reactance for a leading low PF. It must be always noted that overcorrection is also not a good thing as it bring about the same inefficiencies. Here are some of the common methods used by most power utilities for PFC: Static VAR Compensators, Capacitor Banks, Synchronous Condenser, and Static Synchronous Compensator. These methods are discussed in detail below.

2.5.1 Static VAR Compensators

A Static VAR Compensator (SVC) is a high voltage shunt device that uses power electronics to controls power flow and improve transient stability of a power grid [6]. It regulates voltage at its terminals by controlling the amount of reactive power absorbed or injected to the network. It generates reactive power to raise the system voltage and absorbs reactive power to reduce the voltage. This is done through the switching on and off of three-phase capacitor banks and inductor banks.

In general, an SVC is a shunt connected static VAR absorber/generator. This is done through exchanging its output current from inductive to reactive and vice versa in order to maintain specific power system parameters [7]. Its main components include the Thyristor Controlled Reactor (TCR), Thyristor Switched Capacitor (TSC), Harmonic Filters and Mechanically Switched Capacitor. Figure 2.0 shows a SVC connected to a power line.

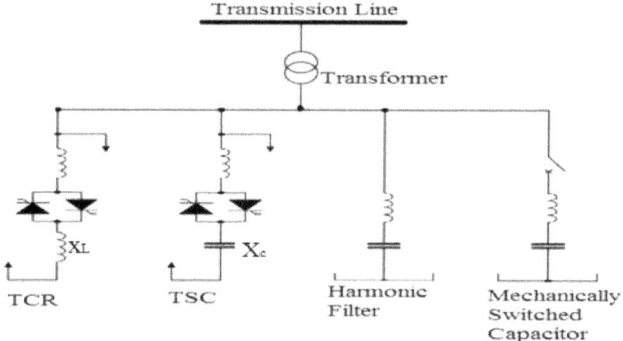

Figure 2.0: One line diagram for a SVC.

The harmonic filter circuit is designed to eliminate harmonics generated by the load and thyristor-controlled circuit. The TSC outputs a stepped response while the TRC cater for a smooth variable susceptance. In steady state, the whole machine will maintain a set parameter by absorbing or injecting reactive power into the network.

The steady state operation of a SVC is governed by the following voltage current relationship:

$$V = V_{ref} + X_{SL}I \ \text{---(2.13)}$$

Where:V and I are the controller's total RMS voltage and current, respectively and V_{ref} is the
reference voltage.

When shown graphically (Figure 2.1), the slope X_{SL} ranges from 2 to 5% with respect to the SVC
base. This is very important to avoid reaching limits in small variations of voltage since
controlled voltages range is within 5% of V_{ref} [8].

The following Matlab simulation illustrates the steady-state and dynamic performance of an SVC
regulating voltage on a 500 kV, 60 Hz, 3000 MVA system. The Static Var Compensator block
models a +200 Mvar/-100 Mvar SVC

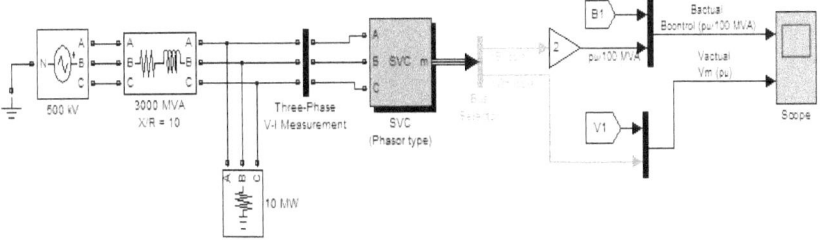

Figure 2.2: Simulation block diagram of a SVC.

The SVC is set to Voltage regulation mode with a reference voltage Vref = 1.0 pu. The voltage
droop reactance is 0.03 p.u./200 MVA, so that the voltage varies from 0.97 p.u. to 1.015 p.u.
when the SVC current goes from fully capacitive to fully inductive. The figure below shows the
SVC V-I characteristic. The Three-Phase Programmable Voltage Source is used to vary the
system voltage and observe the SVC performance. Initially the source is generating its nominal
voltage (500 kV). Then, voltage is successively decreased (0.97 p.u. at t = 0.1 s), increased (1.03
p.u. at t = 0.4 s) and finally returned to nominal voltage (1 p.u. at t = 0.7 s).

The figure below show the bus voltage when there is no SVC connected.\

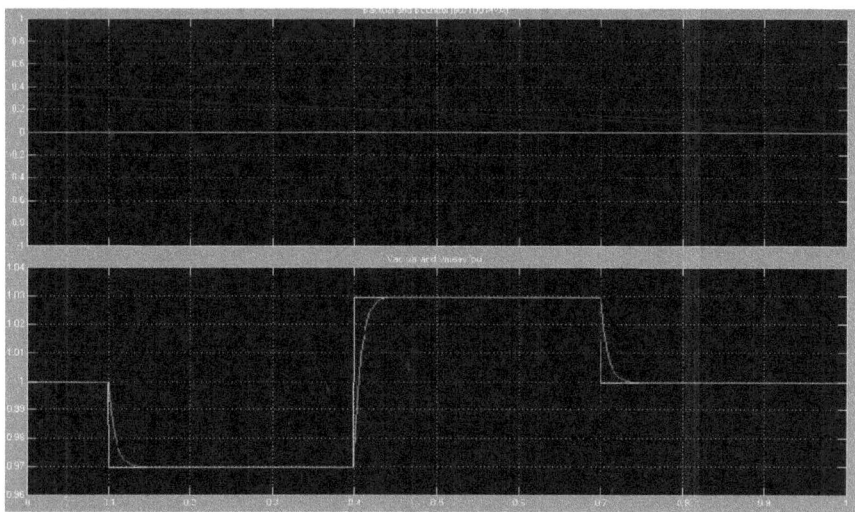

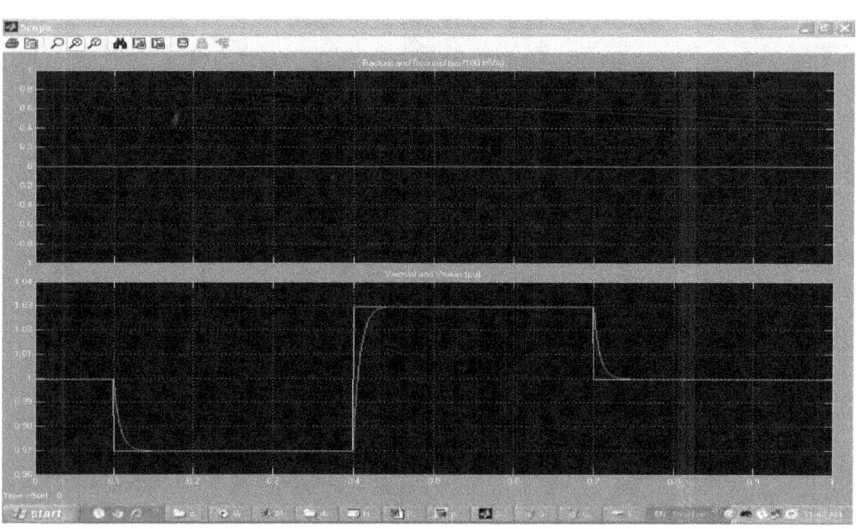

The following figures show the SVC dynamic response to voltage steps on the Scope when SVC is used. Trace 1 shows the actual positive-sequence susceptance B1 and control signal output B of the voltage regulator. Trace 2 shows the actual system positive-sequence voltage V1 and output Vm of the SVC measurement system.

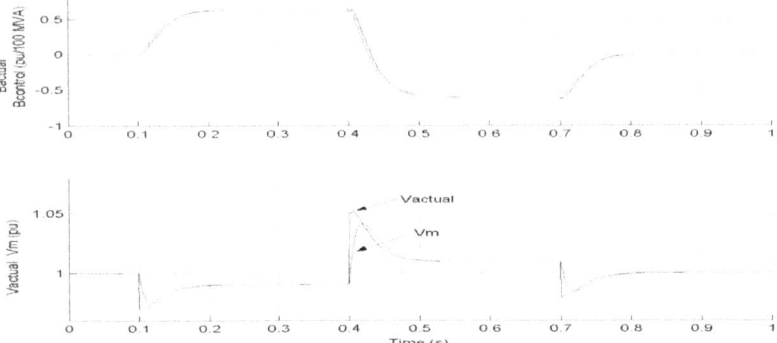

Figure 2.3: Matlab scope output.

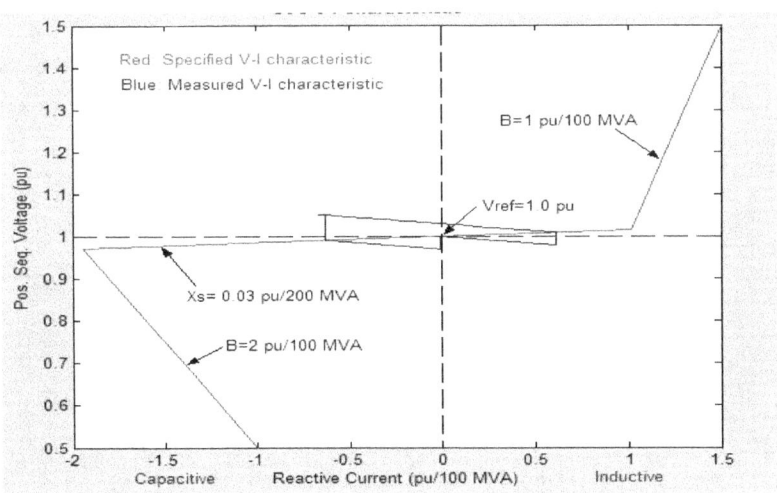

Figure 2.1: Steady state V-I characteristics of a SVC.

2.5.2 Capacitor Banks

These are a set of capacitors connected so as to operate collectively. It is the most common method of PFC. They supply reactive power needed to the load, thus reducing the total current supplied by the power company. Based on the location of the capacitor banks and operation, we can classify the types of compensation into individual, group and centralized compensation [10].

In individual compensation, capacitors are connected in parallel directly to the load's terminals. It is usually applied to transformers, motors and loads with high time of operation. The capacitor bank is switched on simultaneously with the load in which the PF is to be corrected. No additional control is required in this type of compensation. See figure 2.4a below.

Group compensation (Figure 2.4b) involve grouping together the inductive loads and then fitting a common capacitor bank. When compensation is done centrally (Figure 2.4c), an automatic capacitor bank with automatic control is used. Reactive power is subdivided into a number of capacitor steps that can be connected or disconnected independently. The more electrical steps a bank has, the higher the response to changes in reactive power of the network. The controller measures the PF of the system and connect a correct number of capacitors to meet the requirements of the set PF value.

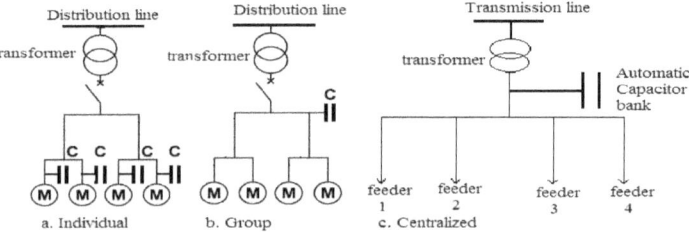

Figure 2.4: Capacitor bank connections.

Example 3: This example shows measurements conducted on a motor feed lines without (and with its nominal load). The measured quantities were apparent power (S_L), line current (I_L) and the power factor (PF).

	Without	4uF	12uF
I_L (A)	1.20 (2.6)	0.91 (2.45)	0.38 (2.24)
S_L (VA)	274 (597)	204 (562)	86 (511)
PF	0.25 (0.85)	0.31 (0.9)	0.73 (0.98)

Table 2.2: Power factor correction example.

Note that the values in brackets are for when the motor was loaded. The behavior was the same i.e. a drop in apparent power and line current as PF was improved. Extremely low PFs were also experienced at no load.

When you want the system to operate at a certain PF and the current PF is known, the following equation (2.13) can be used to determine the required capacitor bank's rating.

$$Q_C = P_{elec}(\tan(\cos^{-1} PF_{old}) - \tan(\cos^{-1} PF_{new})) \text{-----------------} (2.13)$$

Where: P_{elec} = input power in watts with rated output horsepower of a generator. It is calculated as:

$$Pelec = \frac{HP \times 746}{Eff} \text{----------------------------------} (2.14)$$

Eff = Full load efficiency in per unit notation (from datasheet).

PF_{old} = uncorrected full load PF.

PF_{new} = desired power factor.

The installation of shunt capacitors requires that you investigate the following additional factors in order to avoid problems after installation.

2.5.3 Harmonic resonance:

All circuits containing both capacitances and inductances have one or more natrual frequencies. When one of those frequencies lines up with a frequency that being produced on the power system, resonance can develop in which the voltages and current at that frequency continue to persist at high values.

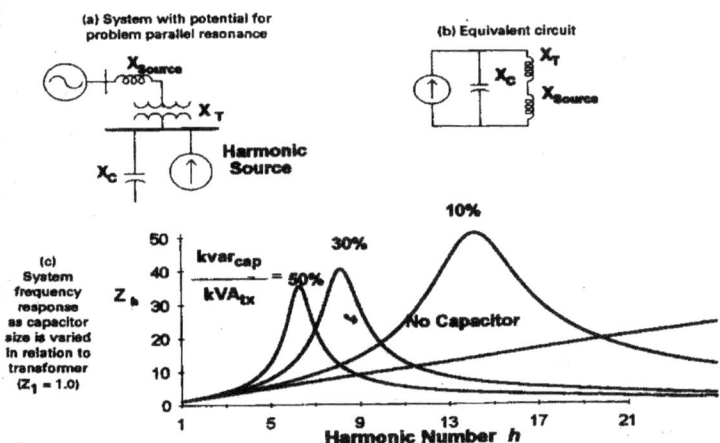

Natural and harmonic impedances of the system are analyzed using frequency scans to determine the effects of voltage amplification. Resonant frequencies are excited during switching events and could cause detrimental voltage amplifications.

Frequency scans are used to analyze the harmonic impedances and the resonant or natural frequencies of the system to determine the voltage amplification effects. The resonant frequencies will be excited during an event, such as switching and could cause voltage amplification at these frequencies. In a frequency scan, a current or voltage of a specific frequency is injected at one bus and the currents and voltages are measured at buses of concern. The frequency of the source is increased in small increments, for instance, 10 Hz and a scan of the system is made at each of these frequencies. Frequencies scans are performed up to the 24th harmonic.

To determine the harmonic impedance of a system as seen from a node, one Ampere of current is injected into the node and the voltage is measured at that node. Since the current is one Ampere, the voltage equals the impedance in magnitude. The current injection across the frequency range studied provides a measure of how the particular system will respond to transient voltages and currents of given frequency content.

All circuits containing both capacitances and inductances have one or more natural frequencies. When one of those frequencies lines up with a frequency that being produced on the power system, resonance can develop in which the voltages and current at that frequency continue to persist at high values. The resonant frequency number (h_1) due to the switched capacitor banks can be calculated using the equation:

$$h_1 = \sqrt{\frac{MVARsc}{MVARc}} \hspace{1cm} \text{-- (2.15)}$$

Where MVARsc is the short circuit rating of the source and MVARc is the rating of the capacitor bank.

The following simulation illustrates the use of the three-phase harmonic filter block to filter harmonic currents generated by a 12-pulse, 1000 MW, ac/dc converter in a 500 kv, 60 Hz system. the filter set is made of the following four components providing a total of 600 MVAR:
one 150 mvar c-type high-pass filter tuned to the 3rd harmonic (f1)
one 150 mvar double-tuned filter tuned to the 11/13th (f2)
one 150 mvar high-pass filter tuned to the 24th (f3)
one 150 mvar capacitor bank.

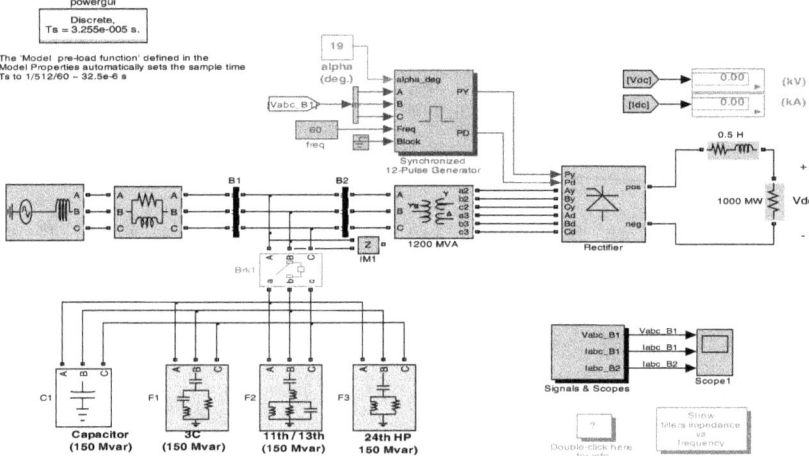

Three-Phase Harmonic Filters Used on a 12-pulse AC/DC Converter

Figure 2.5: Power Harmonic Filters Simulation Model.

- **Case A: When breaker is off i.e. no capacitor and no filter:**

In this condition we do expect to see some harmonics on the system. the currents and voltages waveforms are distorted meaning the system is infested with harmonics. the fast fourier transform (fft) analysis showing the spectrum of the voltage shows the distorted signal, where it is shown that the total harmonic distortion (thd) reaches the heights of 17.78%. the frequency scan shows there is no resonance frequency

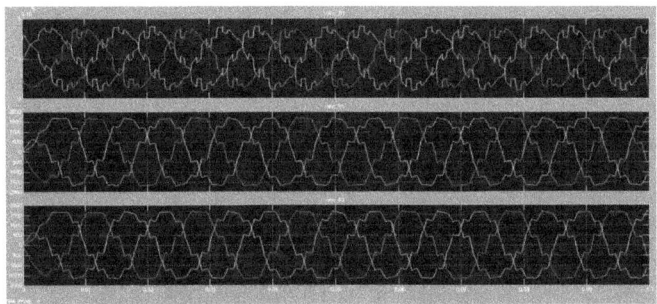

Figure 2.6a: Scope 1 display for case A.

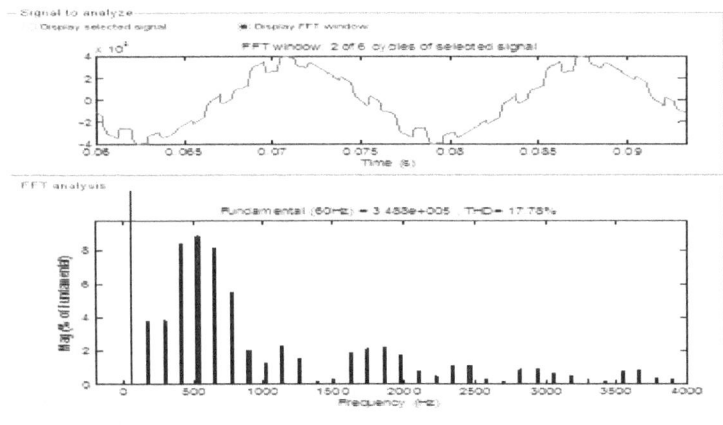

Figure 2.7a: Output of FFT analysis for case A.

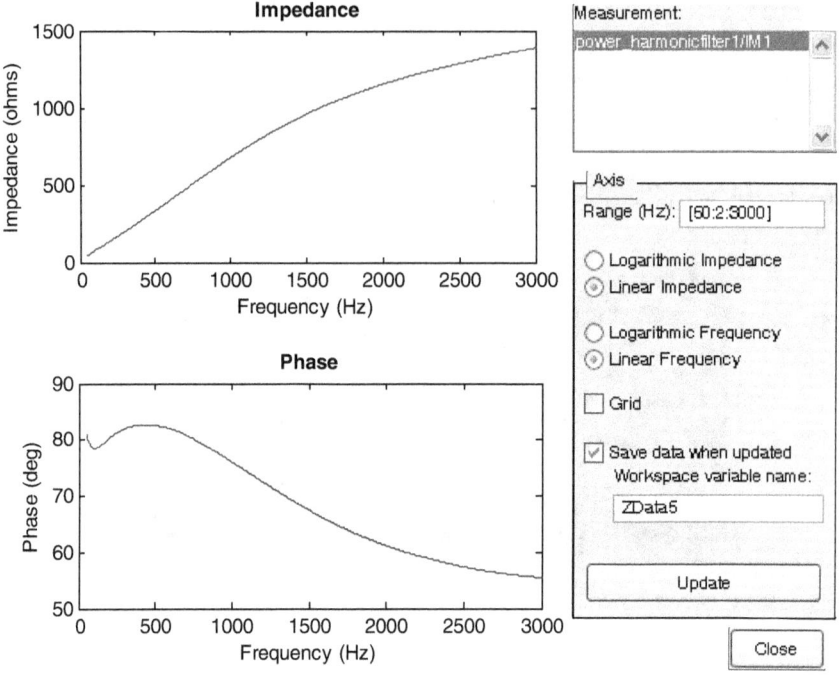

Figure 2.8a: Frequency scan for case A.

- **Case B: When breaker is on but filter removed**

This means that the breaker is on and the filter is removed. waveforms of this conditions from the simulation are shown in the figures to follow.

the waveforms of the current and the voltage show a high level of distortion because the harmonics are not filtered. Also the system shows some parallel resonant frequency around 400 Hz with peak magnitude of input impedance = 2000 ohm. So if there is any harmonic at this frequency it can be amplified and cause distortion on the system. For example, if 1A of 400 Hz harmonic was injected at the point of measuring the impedance, it will cause distorted voltage 2000 V of this harmonic. To mitigate harmonics we use filters, and a system with filters is simulated in the third condition

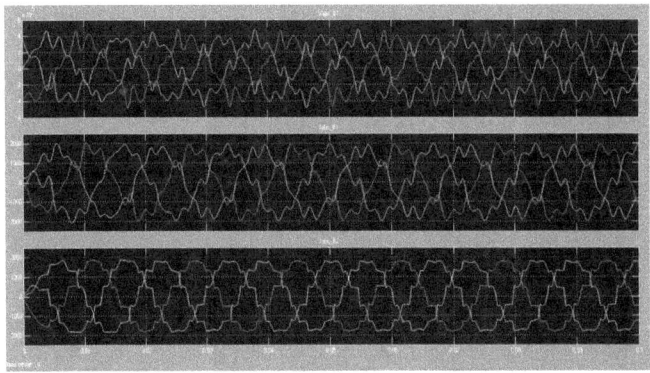

Figure 2.6b: Scope 1 display for case B.

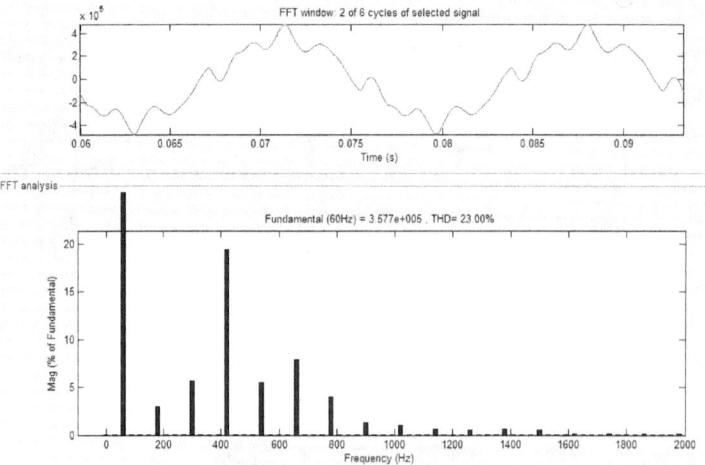

Figure 2.7b: Output of FFT analysis for case B.

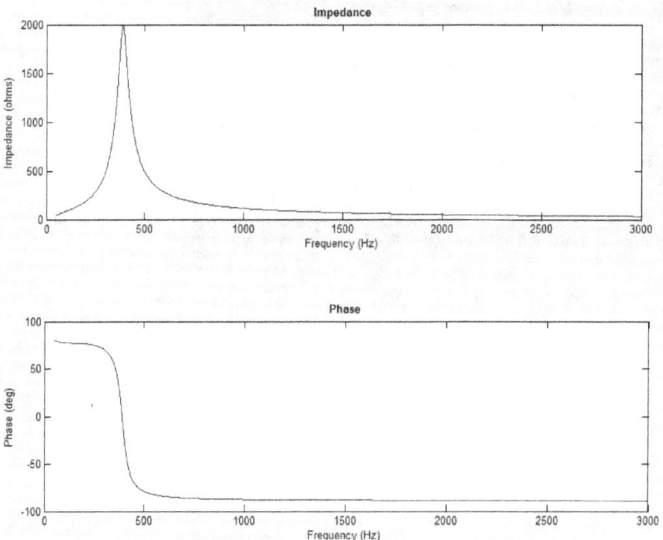

Figure 2.8b: Frequency scan for case B.

- **Case C: When breaker is on and filter is present**

from the waveforms of the voltage and current, look inside scope1. Compare the currents flowing into bus B1 with those flowing into bus B2. You can see that the harmonic filters almost eliminate the harmonics generated by the converter. The harmonic filters reduce the THD of the current injected in the system to 0.7%. From plot the impedance vs. frequency of the system with harmonic filters, there are 4 series resonance frequencies

3rd harmonic frequency: 3*60=180 hz

11 harmonic frequency: 11*60=660 hz

13 harmonic frequency: 13*60=780 hz

24 harmonic frequency: 24*60 = 1440 hz

So the three series resonance filters will shunt the 3rd , 11, 13 and 24 harmonic of the load from the line.

But the filter and capacitor will cause shunt resonance at frequencies 180 Hz, 300 Hz, 670 Hz and 800 Hz, so if there is any harmonic at these frequencies it can be amplified and cause distortion on the system.

The shunt resonance at frequency 800 hz has peak magnitude of input impedance = 400 ohm..

So for example, if 1A of 800 Hz harmonic was injected at the point of measuring the impedance, it will cause distorted voltage 400 V of this harmonic.

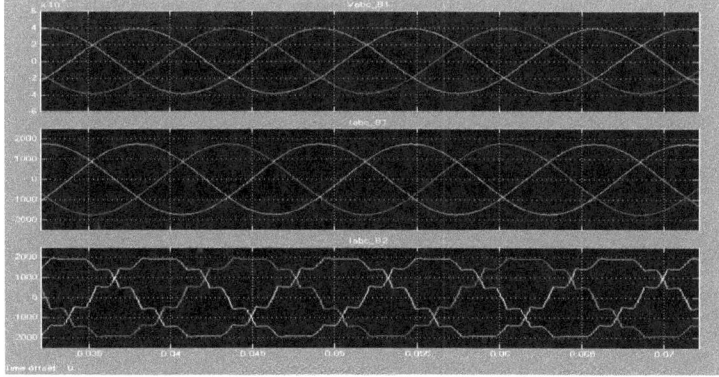

Figure 2.6c: Scope 1 display for case C.

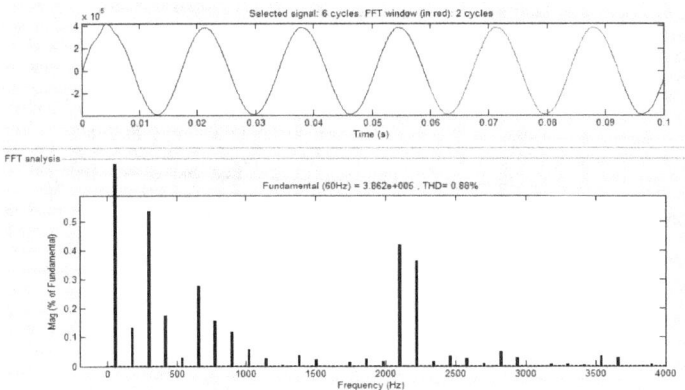

Figure 2.7c: Output of FFT analysis for case C.

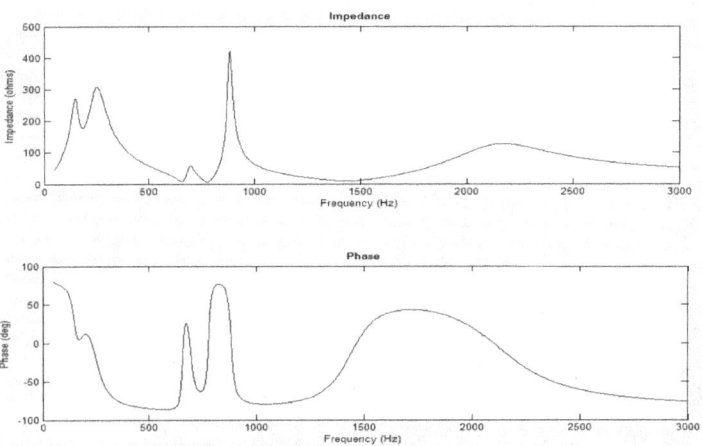

Figure 2.8c: Frequency scan for case C.

Analyzing the curves above, we can see that introducing capacitors to the circuit increases the THD (from 18% to 23% in this case) which can be eliminated by installing filters. Other than reducing the harmonic voltages and currents in the power system installations, AC harmonic shunt filters are also supply the reactive power consumed by the converter. The frequency response of the three cases figure 2.8 a, b and c show that installing capacitors bring about resonance, only one observed resonance frequency in this case around 400Hz. Filtering the capacitors normalized the frequency response.

2.5.4 Voltage magnification

This problem only prevails if capacitor banks are installed at both low voltage and high voltage side of the network. When the natural frequencies at high voltage side and high voltage are equal, voltage magnification will occur. These frequencies are given by

$$f_p = \frac{1}{2\pi\sqrt{L_p C_p}} ; f_s = \frac{1}{2\pi\sqrt{L_s C_s}} \text{-----------------------(2.16)}$$

Where: fs and fp are the high voltage and low voltage natural frequencies

Ls and Lp are the high voltage and low voltage line inductances and

Cs and Cp are the high voltage and low voltage capacitances.

2.5. 5. Sustained over voltages

This is also an important matter when considering the installation of PFC capacitors. When a capacitor bank is energized, the steady state voltage is determined using the following equation.

$$V_{bus} = V_p \left(1 + \frac{Xs}{Xc - Xs} \right) \quad\text{------------------------------------ (2.17)}$$

Where Vp is the bus voltage before energization of capacitors, Xs is the reactance of the source and Xc is the reactance of the capacitor banks.

2.5.6 Switching surge and insulation co-ordination

The ability of transmission equipment and the capacitors to withstand the high switching voltages is also a vital characteristic. Switching surges can last from a few milliseconds to a couple of cycles. Some of the surge withstand standards specified by industry standards are stated in the table below.

Equipment	BIL	kV level (KV)	Surge limit(PU)	IEEE or ANSI Std.
Transformer	900	745	3.9	C57.12
Circuit Br.	900	675	3.6	C37.06
Capacitor	1050	750	3.9	IEEE 824

Table 2.3: Switching Surge Voltage Limitations.

2.5.7 Back to back switching

Back to back switching becomes a problem when energizing a bank step with an adjacent capacitor already in service. Currents of high frequency and magnitude can prevail, thus it must be limited to acceptable levels. This is usually done using a series reactor by-passing the circuit switcher for each step to limit the inrush current. The configuration is shown in the figure below.

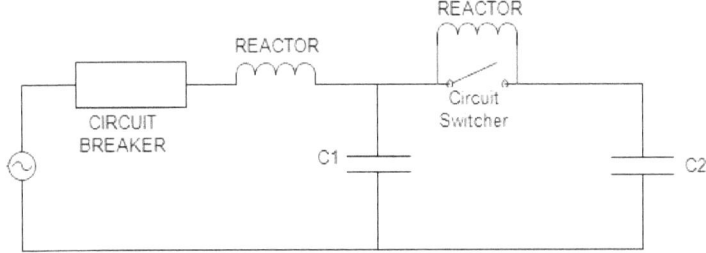

Figure 2.6: Back to back capacitor switching with series reactor.

2.5.8 Synchronous Condenser

A synchronous condenser (sometimes called synchronous compensator) is a synchronous motor that is not attached to a prime mover, only a small pony motor is attached to the synchronous machine input used to accelerate it to synchronous speed [11]. The PF here is controlled by the generated voltage which is a result from DC excitation. In this mode of operation, the synchronous machine acts as a continuous source or sink of reactive power controllable by its field excitation.

The synchronous machine operate at the borderline between motor and generator operation as the active power to/from the machine becomes zero. As the machine becomes over-excited, the power factor becomes zero (cos90=0). With the power factor fixed at zero, increasing the field current raises |E| and the machine current. The current leads terminal voltage V_t. In this way, a high but controllable level of reactive power is produced at zero power factor. Its operation obeys equation 2.15 and can be illustrated using figure 2.7.

$$V_t = E + jI_a X \text{ -- (2.18)}$$

Where Vt is the terminal voltage, E is the internal voltage, Ia is the current and X is the machine reactance.

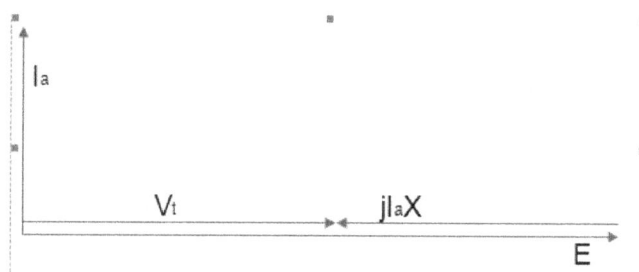

Figure 2.7: Synchronous condenser operation.

Synchronous condensers provide a step less automatic PFC and can produce up to 1.5 times its rated MVARs [12]. They are relatively small in size and produce no switching transients. Other than PFC, they can also absorb some harmonics in the network and also boost it during short term power outages.

2.5.9 Static Synchronous Compensator

A Static Synchronous Compensator (STATCOM) has similar characteristics to the synchronous condenser but superior to it since it is an electronic device, thus it has no inertia and better dynamics. It main functions include voltage control, power oscillation damping, transient stability and reactive power control. The STATCON generates and absorb reactive power purely by means of electronic processing of voltage and current waveforms in a Voltage Source Converter (VSC) [13].

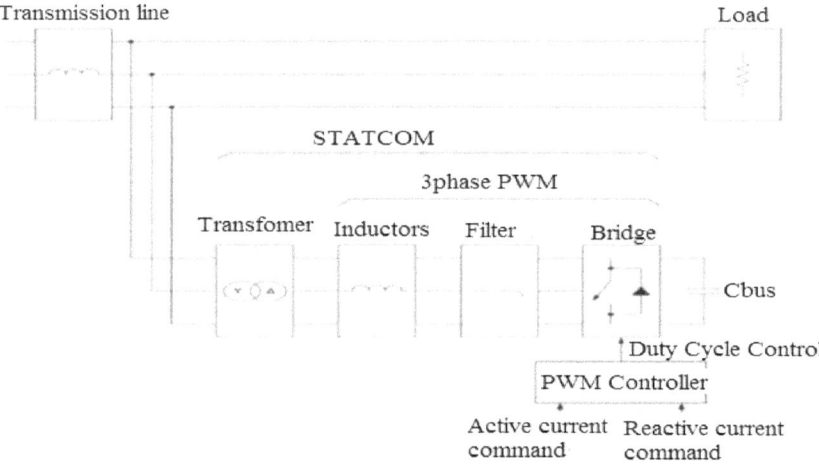

Figure 2.8: STATCOM Block Diagram [13].

Depending on the power rating of the STATCOM, different technologies are used for the power converter. High power STATCOMs (several hundreds of Mvars) normally use GTO-based, square-wave voltage-sourced converters (VSC), while lower power STATCOMs (tens of Mvars) use IGBT-based (or IGCT-based) pulse-width modulation (PWM) VSC.

In the follwing simulation, the STATCOM consists of a three-level 48-pulse inverter and two series-connected 3000 μF capacitors which act as a variable DC voltage source. The variable amplitude 60 Hz voltage produced by the inverter is synthesized from the variable DC voltage which varies around 19.3 kV.

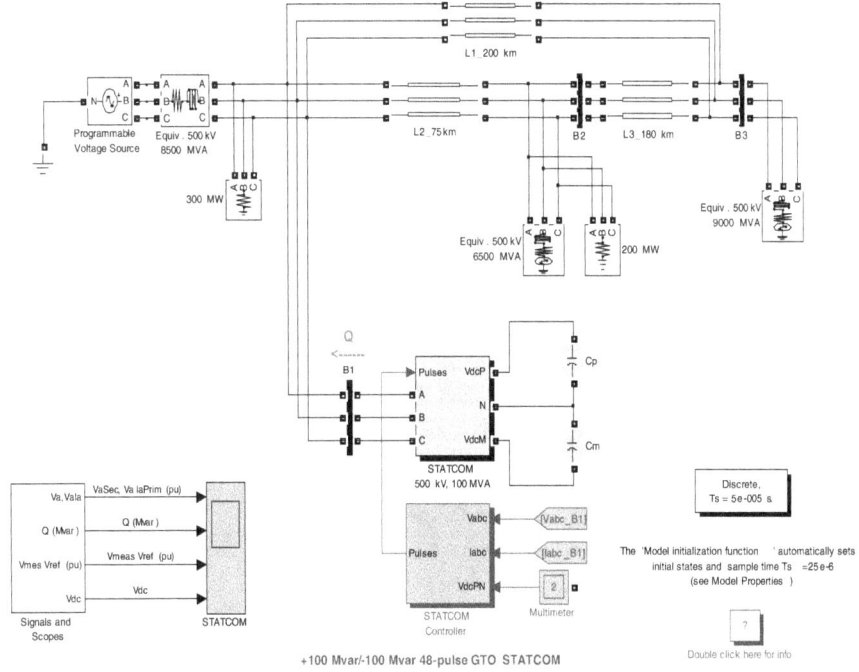

Fig. 7. SPS Model of the 100 Mvar STATCOM on 500 kV Power System

The control system task is to increase or decrease the capacitor DC voltage, so that the generated AC voltage has the correct amplitude for the required reactive power. The control system must also keep the AC generated voltage in phase with the system voltage at the STATCOM connection bus to generate or absorb reactive power only (except for small active power required by transformer and inverter losses).

The control system uses the following modules
1) PLL (phase locked loop) synchronizes GTO pulses to the system voltage and provides a reference angle to the measurement system.

2) Measurement System computes the positive-sequence components of the STATCOM voltage and current, using phase-to-dq transformation and a running-window averaging.

3) Voltage regulation is performed by two PI regulators: from the measured voltage Vmeas and the reference voltage Vref, the Voltage Regulator block (outer loop) computes the reactive current reference Iqref used by the Current Regulator block (inner loop). The output of the current regulator is the α angle which is the phase shift of the inverter voltage with respect to the system voltage. This angle stays very close to zero except during short periods of time, as explained below.

4) A voltage droop is incorporated in the voltage regulation to obtain a V-I characteristics with a slope (0.03 pu/100 MVA in this case). Therefore, when the STATCOM operating point changes from fully capacitive (+100 Mvar) to fully inductive (-100 Mvar) the SVC voltage varies between 1-0.03=0.97 pu and 1+0.03=1.03 pu.

5) Firing Pulses Generator generates pulses for the four inverters from the PLL output (ω.t) and the current regulator output (α angle).

To explain the regulation principle, let us suppose that the system voltage Vmeas becomes lower than the reference voltage Vref. The voltage regulator will then ask for a higher reactive current output (positive Iq= capacitive current). To generate more capacitive reactive power, the current regulator will then increase α phase lag of inverter voltage with respect to system voltage, so that an active power will temporarily flow from AC system to capacitors, thus increasing DC voltage and consequently generating a higher AC voltage.

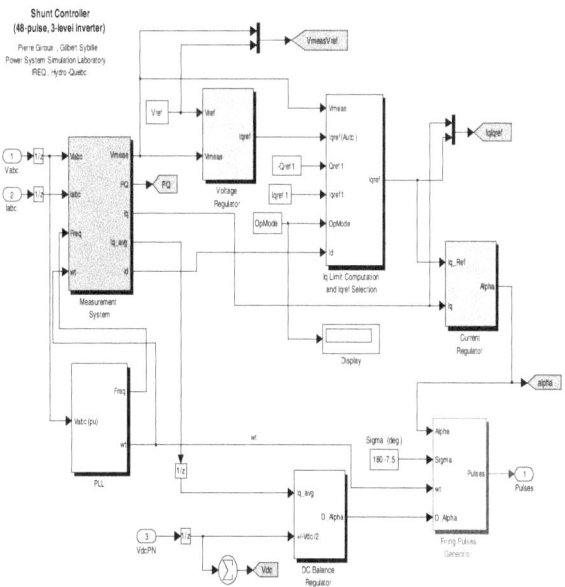

Fig. 8. Shunt Controller (48-pulses, 3 level inverters)

Initially the programmable voltage source is set at 1.0491 pu, resulting in a 1.0 pu voltage at bus B1 when the STATCOM is out of service. As the reference voltage Vref is set to 1.0 pu, the STATCOM is initially floating (zero current). The DC voltage is 19.3 kV. At t=0.1s, voltage is suddenly decreased by 4.5% (0.955 pu of nominal voltage). The STATCOM reacts by generating reactive power (Q=+70 Mvar) to keep voltage at 0.979 pu. The 95% settling time is approximately 47 ms. At this point the DC voltage has increased to 20.4 kV.

Then, at t=0.2 s the source voltage is increased to1.045 pu of its nominal value. The STATCOM reacts by changing its operating point from capacitive to inductive to keep voltage at 1.021 pu. At this point the STATCOM absorbs 72 Mvar and the DC voltage has been lowered to 18.2 kV.

Finally, at t=0.3 s the source voltage in set back to its nominal value and the STATCOM operating point comes back to zero Mvar.

The figure below zooms on two cycles during steady-state operation when the STATCOM is capacitive and when it is inductive.

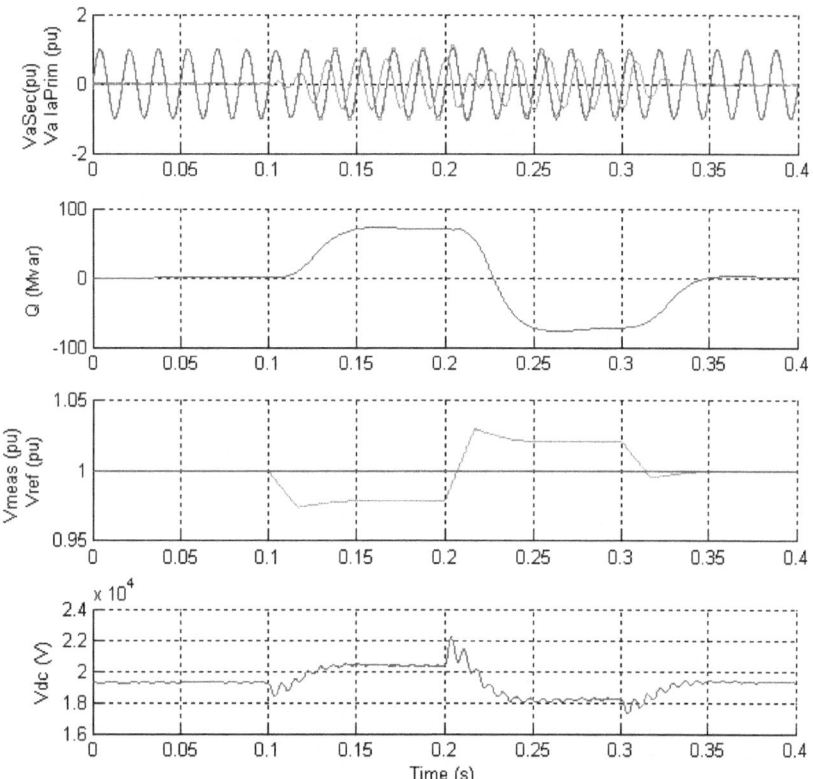

Fig. 9. Waveforms illustrating STATCOM dynamic response to system voltage steps

The following curve (Figure 2.9) shows the steady state behavior of a STATCOM. It exhibits constant current characteristics when the voltage is high/low under/over the limit enabling it to deliver constant reactive power at the limits [8].

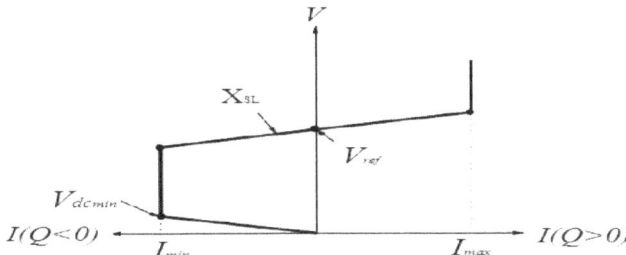

Figure 2.9: Steady state V-I characteristics of a STATCOM [8].

The circuit below (Figure 2.10) shows the Thevenin equivalent circuit of a VSC operating in its switched-mode from which all the equations associated with the power by the VSC and that injected to bus k can be easily derived. It can be represented by equation 2.13.

$$V_{STC} = V_K + Z_{SC}I_{STC}$$ -

(2.19)

Where: V_{STC}, V_K, Z_{SC} and I_{STC} are the voltage at the voltage source inverter, bus k voltage, transformer impedance and current of inverter respectively.

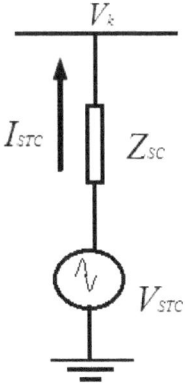

Figure 2.10: STATCOM equivalent circuit.

From figure 2.10, we can deduce that power by the Voltage Source Converter and power injected to bus k are given by equation (2.20) and (2.21) respectively.

$$S_{STC} = V_{STC} \, I_{STC}^{*} = V_{STC}^{2} \, Y_{SC}^{*} - V_{STC} Y_{SC}^{*} V_{K}^{*} - (2.20)$$

$$S_{K} = V_{K} I_{STC}^{*} = V_{STC} Y_{SC} V_{K}^{*} - V_{K}^{2} Y_{SC} \quad - (2.21)$$

Where V_{STC} and I_{STC} are the voltage and current at the voltage source converter, Ysc is the conductance of the STATCON and Vk is the voltage at bus k.

2.6 Reactive Power Planning

Reactive Power Planning (RPP) aims at determining and apportioning the sizes of reactive power resources. Static resources i.e. reactors and capacitors are sized and allocated based on normal operation of the power system. Dynamic resources are allocated and sized to support the system's behavior in case of any single contingency (N-1) situation. In order to optimally solve this problem, various limits should be succeeded [15].

2.6.1 Static Reactive Resource Sizing and Allocation

These affect the voltage stability and the voltage profile. Four aims associated with optimization of static reactive power resources namely voltage stability, voltage profile, system losses and cost of the installation are explained below.

a) Voltage Profile – This refers to the acceptable range of voltage magnitudes, usually from 0.95 to 1.05pu. The normal voltage magnitude is 1pu. In practical situations it is difficult to keep the voltage fixed at 1pu especially for load (PQ) buses. The voltage at a generator bus is set by the operator. An index (P_{prof}) for determining an acceptable voltage profile is calculated using the equation below.

$$P_{prof} = \sum_{i=1}^{N} \left(V_i - V_i^{set}\right)^2 \text{-----------------------------------(2.22)}$$

Where V_i is the voltage at bus i, V_i^{set} is the reference voltage of bus i and N is the total number of buses. Since a desired voltage profile is one with all the buses at 1.0pu, therefore an ideal voltage profile is zero. This means that a low voltage profile is desired in RPP.

b) Voltage Stability – This refers to the response of the system voltages to an increase in the load. The voltage stability index (P_{stab}) is the maximum amount of reactive power the weakest bus of the system can handle before its voltage becomes unstable (i.e. ACLF diverges). A high voltage stability index is favorable. To find Pstab for the whole system, the reactive power loads of all buses proportionally increased until nose point is reached for the weakest bus. The total increase in reactive power is considered as Pstab [15]. The figure below shows the curve for voltage stability.

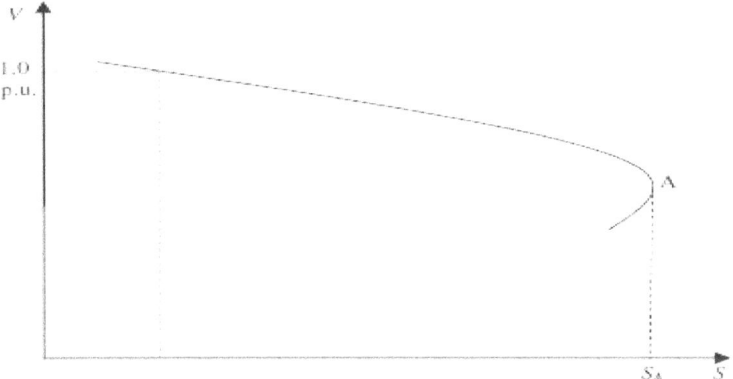

Figure 2.11: S-V curve [15].

If load is greater than S_A the system becomes voltage unstable.

c) System losses – This can be expressed using the following equation.

$$P_{loss} = \sum_{i=1}^{N} g_i \left[\left(V_i^s \right)^2 + \left(V_i^r \right)^2 - 2V_i^s V_i^r \cos\theta_i \right] \text{------------------------(2.23)}$$

Where g_i is the conductance of line i, V_i^s and V_i^r are the sending and receiving end voltages of line i, θ_i phase angle difference of line i and N is the total number of lines. Active losses are desired to be minimal i.e. $V_i^s = V_i^r$ and $\cos\theta_i = 1$.

c) Cost of installation – An optimal installation of reactive power resources should be obtained at the lowest possible cost. The cost index (P_{cost}) is calculated using the following equation.

$$P_{cost} = \sum_{i=1}^{Nc} \left(C_{fi} + C_{vi} Q_i \right) \text{--(2.24)}$$

Where C_{fi} = the fixed installation cost of reactive power resource at bus i, C_{vi} = variable cost of reactive power resource (per kVAR) at bus i, Q_i is the capacity of reactive power resource at bus i and Nc is the total number of allocation points.

Therefore the resulting multi-objective optimization problem is described as:

    i. Minimum voltage profile index (Min. P_{prof})

    ii. Maximum Voltage stability index (Max. P_{stab})

    iii. Minimum active losses (Min. P_{loss})

    iv. Minimum Reactive power resources cost (Min. P_{cost})

Subject to H=0 (load flow equations) and G<=0 (inequality constraints such as limits on active (reactive) power generations of power plants, voltage magnitudes, etc.) may be solved by an optimization method.

2.6.2 Dynamic Reactive Resource Sizing and Allocation

It has been said that dynamic reactive compensators (RPCs) are employed to improve voltage security of the network in reaction to N-1 contingencies of transmission equipment. A system said to be secure if load flow converges and all its voltages are within the acceptable levels i.e. 0.95–1.05 p.u. When reacting to an N-1 contingency, the following conditions may occur (assuming all static reactive resources allocated in the above section are in service):

a) System operate at acceptable voltage levels and load flow converges. In this case no RPC is required.

b) An islanding condition cause load flow not to converge after the contingency. This case cannot be solved by installing RPCs.

c) No islanding condition but load flow doesn't converge. Action is required to solve this situation.

d) Though load flow converges, some voltages happen to lie out of the stipulated range i.e. 0.95 to 1.05 p.u. Installation of RPCs can also eliminate this problem.

Optimal location and sizing of RPCs can help in solving problems associated with conditions c) and d) above.

2.6.3 Solution Procedure:

We may use one of existing powerful optimization algorithms to solve the above optimization problems. If the system under study is small and the search space is limited in terms of the resources candidates, we may search the entire space and calculate the evaluation function; in order to find the optimum solution. If the search space is large, a powerful metaheuristic approach is Genetic Algorithm (GA) by which the solution may be found; quickly and in an efficient manner. An alternative, yet simple solution procedure is depicted in Fig. 10.8. This is called the sensitivity approach in which the evaluation function is initially calculated for the base case. Following that, a small reactive resource is applied at each bus, one-by-one and the evaluation function, recalculated. Based on the resulting calculations, the most sensitive buses are determined. Thereafter, a small reactive resource (say 0.1 p.u. of capacitor) is applied at the most sensitive bus and the whole procedure is repeated. For instance, in the second run, the first bus may be still the most sensitive and a second 0.1 p.u. resource may be added to that bus. The procedure is repeated until no further bus may be found which results in improving the evaluation function. The proposed approach may be used for both small and large systems [15].

The sensitivity approach is summarized in the following flow diagram.

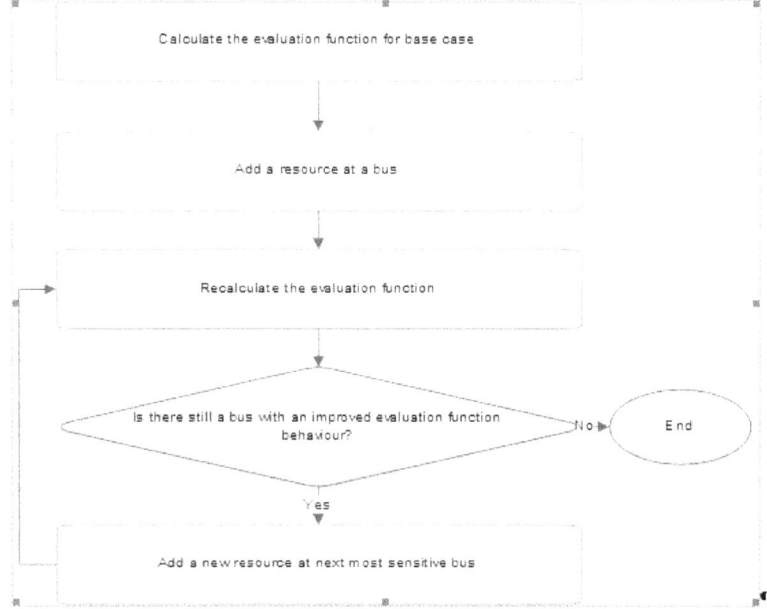

Figure 2.12: Sensitivity approach flow diagram.

2.7 Conclusion

It is vital that the utility company ensures that correct power factor correction equipment are installed where necessary and that violations associated with operating at a low PF are dealt with accordingly. Doing a detailed study of the load behavior before finalizing on the PFC equipment to use is very important and it is the only key towards achieving an effective power factor correction installation.

CHAPTER 3: RESEARCH METHODOLOGY

3.1 Introduction

In this chapter the research methodology used in the study is described. The area where the study was conducted and the design of the study are described. The method used in the collection of data is also described.

3.2 Research Approach and Design

A qualitative approach was followed. This approach helped in analyzing complex load behavior and choosing the optimum method of PFC for each studied substation. The first step of this project was to do an extensive literature review on power factor correction. The main objective reading theory was to answer these three questions:

- What causes Power Factor to become low?
- What impact does low Power Factor have on the transmission network?
- What methods/techniques have been proposed to raise the low Power Factor?

These questions were all answered in several books, online documents and reports on previous work pertaining power factor correction. The second step involved understanding a method used at Swaziland Electricity Company and analyzing it effectiveness based on the results after implementation of PFC.

3.3 Research Setting:

This study was conducted at Swaziland Electricity Company (SEC). Special attention was paid to the Eastern side of its network since it was where extreme conditions were experienced due to the irrigation pumps by the agricultural industries.

3.4 Data Collection

Data pertaining how the system load behave for a period of a year was obtained from SEC's metering department. It was in the form of spreadsheets for daily loading averaged every thirty minutes by Enermax meters in each substation. The load parameters which were recorded are active, reactive and apparent power. These were enough for the calculation of the PF and rating for PFC equipment. After the installation, data was again collected using the same metering equipment.

3.5 Data Analysis

After the data was collected it was organized and analyzed. This was done before and after implementation of power factor correction in the network. Tables and graphs were drawn to aid in the analysis. Analyzing also involved looking closely at the loads involved to help in explaining how it influenced the results. A computer program (DigSilent Power Factory) was used to simulate the system's PFC devices behavior when load properties are varied.

3.6 Research Rationale

As stated above, a qualitative research method approach was chosen for this project. Using a qualitative simulation design and a qualitative research methods helped in determining an optimal solution. The research sample was chosen from the eastern part of the grid for a compact simulation design and also because it is where the system PF is at its lowest. These substation had their loads studied closely before a decision about PFC sizing chosen. It is very difficult if not impossible to determine PFC equipment sizing accurately for each substation without qualitatively studying its load behavior. The simulation design was designed to check if the installation equipment they would really enhance the power system's performance.

CHAPTER 4: DATA COLLECTION

4.1 Introduction

In this chapter loading data from several substations of SEC obtained from the metering department were presented in graphs and tables. Five substations (Big bend, Maloma, Mhlume, Ngomane and Tabankulu) were picked from the list and then their load behavior were analyzed individually.

4.2 SEC's loading data

The following table (Table 4.1) shows a summary of loading information in each of SEC's substations. These readings represent the total load of the feeders and was recorded in the 11kV busbars of each substation. The minimum and maximum readings of the apparent power (measured in KVA) with their corresponding power factors were obtained from Enermax tables. Zero, suspect and gap values were omitted as they would have portrayed a wrong image about the load behavior. These depicted complete shutdown of the loads and the main reason could be for installation of new equipment at the substation and maintenance of substation equipment.

The PF was often lower when the load was drawing less power from the network compared to maximum loading. This is because for most inductive loads the PF is in its worst conditions at minimum loading and highest at full load. Most motor loads have PF varying from 0.3 to 0.8 lagging. There were a few cases where the above phenomenon wasn't obeyed (Endzingeni, Nkhaba and Tabankulu substations).

Substation loading information				
Substation	S_{min}	PF_1	S_{max}	PF_2
Balegane	2.59	0.37	1374.53	0.88
Big Bend	202.40	0.86	9520.75	0.89
Dvokolwako	2.59	0.37	1374.53	0.88
Endzingeni	0.16	0.00	306.06	0.32
Ezulwini	50.41	0.84	3417.98	0.96
Hlatikhulu	49.21	0.94	3200.51	0.98
Illovo	0.04	1.00	14036.40	0.99
Kalanga	3.33	0.20	6210.34	0.94
Kent Rock	1750.89	0.90	13885.73	0.94
Lawuba	6.34	0.87	2241.39	0.98
Lobamba	239.72	0.82	10882.76	0.94
Magwabayi	2488.43	0.77	11565.00	0.89
Malkerns	0.08	0.24	6443.98	0.93
Maloma	1.07	0.45	4017.95	0.63
Mankayane	4.43	0.76	3126.32	0.96
Manzini North	672.33	0.84	12921.84	0.97
Manzini	93.99	0.87	15674.58	0.95
Matsapha 1	1.36	0.71	6878.66	0.94
Mayiwane	20.21	0.74	1386.80	0.93
Mhlambanyatsi	3.96	0.92	1313.98	0.98
Mhlume	186.63	0.60	8896.94	0.82
Mnkinkomo	1347.18	0.93	22712.40	0.88
Mpaka	11.91	0.81	1397.27	0.95
Mpisi	55.74	0.85	3447.31	0.93
Ncandvweni	0.11	0.71	241.44	0.88
Ngomane	44.21	0.93	9025.52	0.91
Nhlangano 1	17.66	0.73	7342.37	0.95
Nhlanganon 2	3.26	0.20	6476.85	0.82
Nkhaba	3.37	1.00	984.65	0.98
Nkoyoyo	107.99	0.78	8325.36	0.98
Nsoko	63.29	0.67	5600.13	0.85
Old Ngwenya	9.51	0.67	4510.53	0.92
Piggs Peak	3.10	0.34	8020.39	0.91
Pine Valley	5.90	0.89	3420.14	0.96
River Bank	4.19	0.58	9523.75	0.90
Sappi Usuthu	1.22	0.20	2686.00	0.96
Sidvokodvo	114.04	0.70	2233.06	0.83
Sihhoye	16.87	0.64	5974.77	0.84
Sikhuphe	52.80	0.00	1119.53	0.95
Simunye	24.61	0.63	14407.03	0.81
Siphocosini	1.07	0.45	2726.45	0.95
Sithobela	2.31	0.83	2909.76	0.86
Stonehenge	206.92	0.94	12208.89	0.97
Swazi Paper Mills	2.72	0.71	3059.61	0.92
Tabankulu	34.00	1.00	6553.54	0.96
Thomson	76.83	0.85	9399.12	0.84
TOTAL	7990.96	0.68	292982.88	0.90

Table 4.1: SEC loading data.

4.3 Loading analysis for selected substations

The five selected substations are all located in the eastern side of the network. They were selected because of their location. Their loads are industrial and mainly it is where the system's PF is at its worst. The nature of the industrial load is the main reason behind the low PF.

4.3.1 Big Bend Substation

Big Bend substation's load consist of mainly farmers and a sugar cane processing firm. Almost all the stages in the production of sugar depend on electricity as a source of energy. This mean that power is drawn by irrigation pumps, boilers and motors. Figure 4.0 shows how the above loads affected the load profile at the substation for a period of one week (Wednesday 14 November 2012 to Tuesday 20 November 2012).

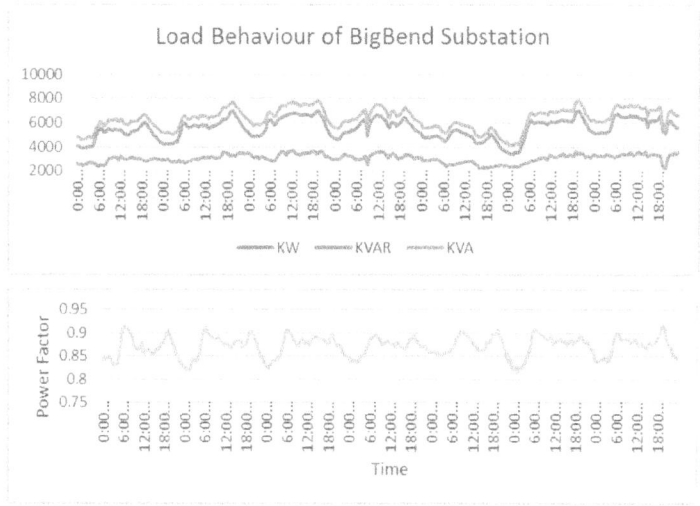

Figure 4.0: Load profile for Big Bend substation (14-20Nov 2012).
Source: SEC Enermax metering tables.

From the curves:

The lowest KVA reading for each day occurred during evening off-peak hours. This is because most of the industrial equipment is being switched off during these times as most of the production takes place during the day.

The real power followed the same trend as the apparent power while the reactive power is varying smoothly between 2.2 and 3.8 MVAR in both off-peak and peak periods. This is probably because reactive loads are present during both day and night i.e. motors during the day or irrigation pumps at night.

The PF was lowest at midnight reaching its maximum at working hours for each day. Reactive loads experience low PF at minimum loading condition and the best when operating at full capacity.

4.3.2 Maloma Substation

Maloma substation's main load is a coal mine. In a mine electricity is used in a number of electrical equipment which dust collectors, lighting equipment, air conditioners, motors etc. These have a high requirement of reactive power as shown in figure 4.3 (Appendix A). We can see that in some cases the value of KVAR absorbed exceeded the KW dissipated. The power factor of Maloma was the worst of the five substations as its power factor was always below 80 percent. The whole load profile is shown in Appendix A.

4.3.3 Mhlume Substation

Mhlume supplies a sugar company and a number of independent sugar cane farmers. Since Mhlume is located in the lowveld of the country, the sugar cane requires extensive irrigation through the use of irrigation pumps controlled by VSDs. Moreover, during processing rotating knives and hummers are used to chop and crush the cane. Most of the equipment used in these production steps include motors in their design, thus they require both reactive and active power in their operation. This can be confirmed from the load profile in figure 4.4 (Appendix A) as its power averaged around 0.85 in this chosen week. There are two specific times where abnormal behavior was noticed: on the 15th just before midnight and on the 16th just after midnight. Here extremely low KW, KVAR, KVA and PF readings were seen and these were considered suspect.

4.3.4 Ngomane Substation

Ngomane is another substation supplying large scale irrigation farms of sugar cane. The load behavior in figure 4.5 (Appendix A) is showing a periodic behavior meaning that they have an irrigation routine with most of the irrigation taking place during the day. Power factor followed the same behavior reaching a maximum of 0.93 at full load and a minimum of 0.87 during light loading.

4.3.5 Tabankulu Substation

Tabankulu estates is an agri-business owned by Tongaat Hulett. It supplies Royal Swaziland Sugar Corporation with around 62'000 tons of sugar annually from its 3767 hectares of fully irrigated cane. The main loads in Tabankulu substation are irrigation pump motors. These are usually controlled by variable speed drives which also increase the reactive power requirement of the load. Although the reactive power requirements were bad as it ranged from 300 to 1900KVAR, the power factor was not that bad like in the other substations as it was always close to 0.95 in this chosen period. The load profile is shown in figure 4.4 (Appendix A).

4.4 Summary

Bigbend, Maloma, Mhlume, Ngomane and Tabankulu substations are all mainly supplying industrial loads and the data collected in each substation resembled the overall load behavior i.e. summed up the contributions by each individual load. Their loads were mostly motors, irrigation pumps and lighting equipment and they experienced low PF at minimum load conditions and the maximum load occurred during working hours as expected.

CHAPTER 5: RESULTS AND ANALYSIS OF THE SIMULATION MODEL

5.1 Introduction

In this chapter we discuss the simulation model of the PFC for a selected part of SEC's network. The load will be varied and particular attention will be paid on the behavior of other circuit parameters. This simulation was carried out in DigSilent power factory where all the simulation results were captured. DigSilent power factory is a very accurate power simulation software which allows you to model any desired power network. This software can run on any machine with windows xp, vista, 7 and 8. Figure 5.0 below shows a snap shot of the simulation environment. The main aim was to study the response of other transmission variables to a change in power factor and reactive power.

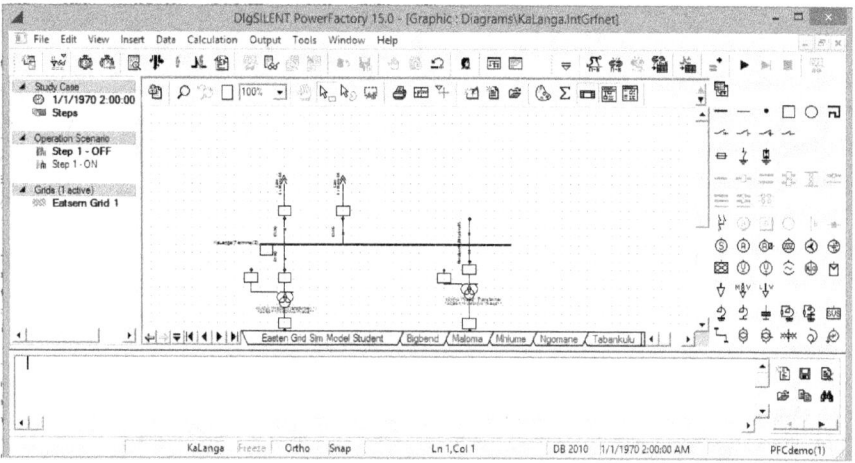

Figure 5.0: DigSilent Power Factory's graphic user interphase.

When fully licensed, DigSilent can perform balanced or unbalanced load flow calculations, contingency analysis, short circuit analysis, harmonics/Power quality, generation adequacy, optimal capacitor placement, cable sizing, motor starting, techno-economical calculations, and RMS/EMT simulations.

5.2 Power Factor Correction Simulation Model

In this model, we first design the simulation circuit with the help of the countrywide transmission diagram. All the selected substations were vital in the simulation. Circuit components like transformers and line conductors were modelled to be as close as possible to those in the actual network. The load behavior was also varied according to the actual data which was captured in previous encounters in the network. This will ensure that the results of the simulation bring about a solution which can be even applied in a real life situation. The figure below shows how the simulation network was modelled.

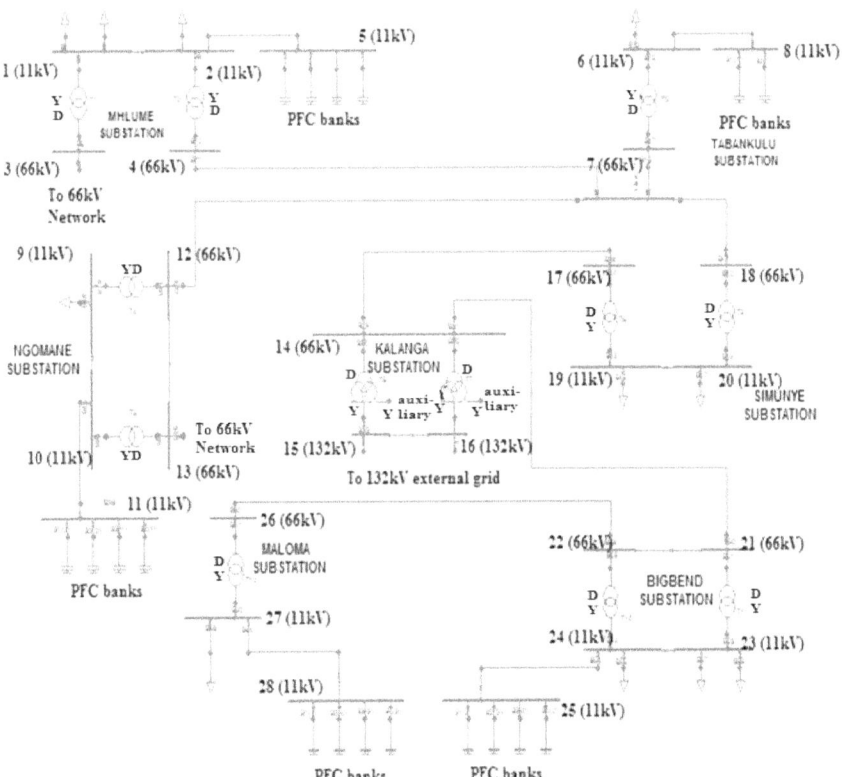

Figure 5.1: The Simulation Network.

Where: 66kV busbars are 3, 4, 7, 12, 14, 13, 17, 18, 21, 22 and 26

11kV busbars are 1, 2, 5, 6, 8, 9, 10, 11, 19, 20, 23, 24, 25, 27 and 28

132kV busbars are 15 and 16

The PFC busbars are 5, 8, 11, 25 and 28 (11kV busbars).

The power source was a 132kV 40MVA external grid.

Figure 5.2 shows how the block diagram of the simulation network with an external network being the 132kV power source and connected to KaLanga substation.

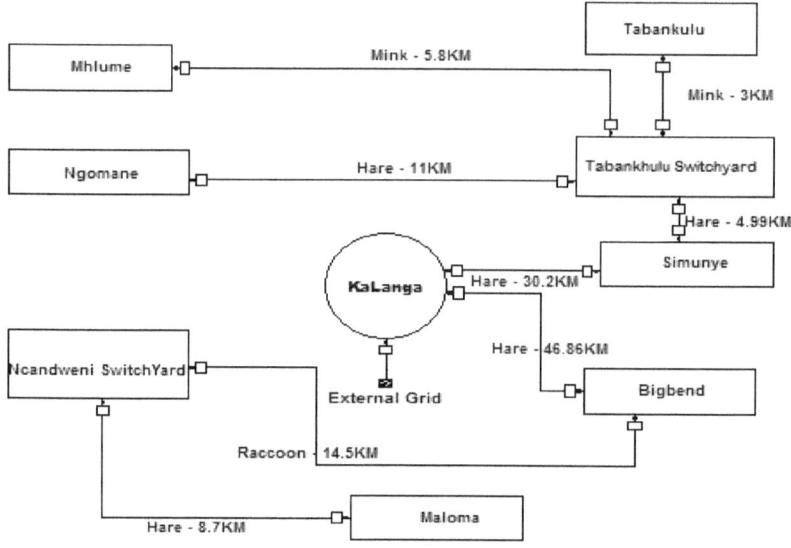

Figure 5.2: Block diagram of the simulation network.

Note: Each rectangular block represents a substation. To view the detailed circuit of each of the selected substations you can refer to Appendix B.

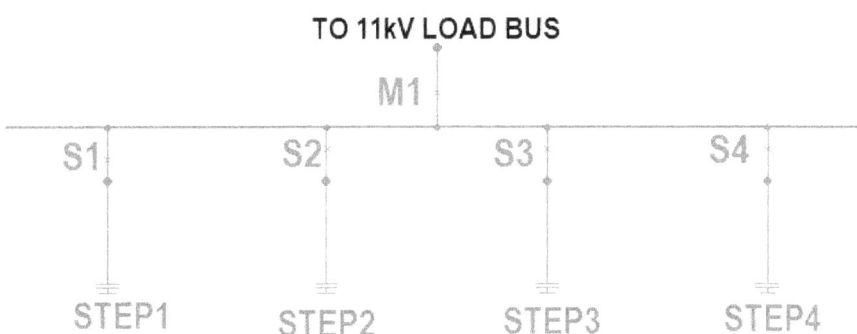

Figure 5.3: Connection of the PFC banks.

The switches S1, S2, S3 and S4 are operated by a PFC controller in response to variations in reactive power requirements of the grid.

5.3 Circuit Components

The main circuit parameters which were modelled to detailed specifications are the line conductors, transformers and tower. These components were named accordingly for easy data capturing and exportation of results for analysis.

5.3.1 Line Conductors

There were three types of conductors used for linking the substations in the selected part of the network. These were the Mink, Hare and Pigeon conductors. Their details are specified in the following table.

Line type	Rated Voltage	Rated Current	Nominal Frequency	R' (20°C)	Overall Diameter
	kV	kA	Hz	Ω/KM	Mm
Hare	66	0.36	50	0.2733	14.12
Mink	66	0.26	50	0.4546	10.98
Raccoon	66	0.30	50	0.3640	12.27

Table 5.0: Three phase conductor's specifications.

5.3.2 Transformers

Only 66/11KV transformers were present in each of the substations. The main difference was the number of transformers in each and the MVA rating of each. Their specifications are given in the following table

Substation	Quantity	Trans MVA rating	Vector type	Taps(Nom)	Impedance (p.u.)
Bigbend	2	20	Dyn11	17(5)	0.101
Maloma	1	5	Dyn11	17(5)	0.073

Mhlume	2	10	Dyn11	5(3)	0.091
Ngomane	2	7.5	Dyn11	17(5)	0.076
Tabankulu	1	10	Dyn11	5(3)	0.091

Table 5.1: 66/11KV Transformers in selected substations.

The other two substations which were also part of the simulation i.e. Simunye substation uses two 66/11kV 10MVA transformers while KaLanga substation is with two 132/66/11kV (3 windings) 40MVA transformers.

5.3.3 Tower

The transmission tower used in the 66KV was based on the geometry specified in figure 5.3.

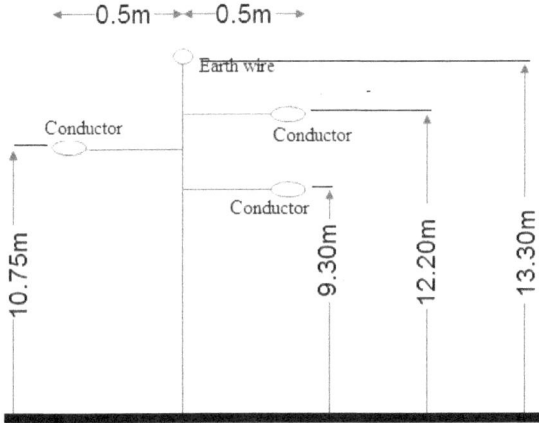

Figure 5.4: SEC Tower G68/79-B configuration.

5.4 Simulation Process

The simulation was based on the Capacitor bank PFC method. Centralized correction was used as the banks we located on 11KV buses at the substations. The PFC controller at each location was modelled to constantly measure the amount of inductive reactive power in the network and switch on capacitor steps to help reduce the magnitude of the reactive power. This was achieved through the use of DigSilent study cases where each case represented the response of the controller to certain load conditions.

The loading was based on the data below which was collected during the periods stated in the following table. Table 5.2 shows how the data was grouped on frequency in steps of 500kVAR. Each unit in the frequency represent a 30 minute average for the inductive reactive power reported in the system in the specified period of time.

Substation	FREQUENCY OF OCCURRENCE						Period	
	500-1000 kVAR	1000-1500 kVAR	1500-2000 kVAR	2000-2500 kVAR	2500-3000 kVAR	3000-3500 kVAR	Start date	Last day
Big bend (Bus 24/25)	11	587	3112	3275	3806	2663	10-Oct-2012	09-Aug-2013
Maloma (Bus 27)	134	4232	4498	2840	3028	9	11-Oct-2012	29-Aug-2013
Mhlume (Bus 1/2)	3698	4041	2368	2041	1703	1115	14-Aug-2012	27-Aug-2013
Ngomane (Bus 9/10)	3242	2881	2879	2155	1438	1288	16-Aug-2012	10-Sept-2013
Tabankulu	4475	2049	1124	0	0	0	15-	10-

Table 5.2: KVAR consumed in selected substations.

The table above show the total KVAR loading in each substation's 11kV busbars i.e. summation of all the 11kV feeders' KVAR. From the table, we can picture out three details which are very useful in choosing the rating and number of steps required in capacitor banks PFC method. These include the total kVAR rating for banks, the number of steps needed and the size of each steps. The amount of absorbed reactive power generated in normal operation in each were selected as follows:

- Big Bend 1500 to 3500 kVAR -4 steps
- Maloma 1000 to 3000kVAR – 4steps
- Mhlume 500 to 2500kVAR – 4steps
- Ngomane 500 to2500 kVAR – 4steps
- Tabankulu 500 to 1500kVAR – 2steps

Frequencies below 2000 were considered less frequent as it meant they only rarely occurred, so their effect on the network was insignificant. Since the maximum number of steps is four, therefore also four study cases were used in the simulation.

The following table shows the data used in each simulation case. Four operation scenarios were examined in each substation except for Tabankulu (two scenarios). Each case will demonstrate how the PFC equipment responds to the amount of KVAR in the line and improving the PF in the process.

Substation	Study Case	KW	KVAR	KVA	Voltage(p.u.)	PF
Bigbend (Bus 24/25)	1	1755	1506	2313	1.00	0.76
	2	2668	2166	3437	0.98	0.78
	3	3977	2791	4859	0.96	0.82
	4	5061	3200	5988	0.93	0.85
Maloma	1	686	1215	1395	0.97	0.49

(Bus 27)						
	2	1202	1583	1988	0.94	0.60
	3	1026	2358	2572	0.89	0.40
	4	2459	2772	3705	0.84	0.66
Mhlume (Bus 1/2)	1	583	551	802	1.00	0.73
	2	1202	1102	1631	0.97	0.74
	3	2067	1628	2631	0.95	0.79
	4	2843	2361	3696	0.92	0.77
Ngomane (Bus 9/10)	1	853	654	1075	1.00	0.79
	2	1712	1135	2054	0.98	0.83
	3	2887	1639	3320	0.96	0.87
	4	4198	2201	4740	0.94	0.89
Tabankulu (Bus 6)	1	723	629	958	1.00	0.75
	2	1979	1192	2310	0.95	0.86

Table 5.3: Simulation loads for the selected substations.

Source: SEC Enermax metering tables (KW, KVAR, KVA and PF).

The loading was applied on the 11kV buses of each substation and the bus for PFC capacitors was also connected to the each substation's 11kV bus (Bigbend (bus 25), Maloma (bus 28), Mhlume (bus 5), Ngomane (bus 11) and Tabankulu (bus 8)).

The capacitors were connected in the following manner. For:

> Case 1: Single step capacitor at each bus (5, 8, 11, 25 and 28)
>
> Case 2: Two capacitor steps at each bus (5, 8, 11, 25 and 28)
>
> Case 3: Three capacitor steps at each bus (5, 11, 25 and 28)
>
> Case 4: Four capacitor steps at each bus (5, 11, 25 and 28)

NB: For bus 8 the 3rd and 4th steps are not included as per its reactive power requirement.

5.5 Simulation results

The simulation results were captured from the output of the load flow calculation. It was based on a balanced positive sequence and the transformers were set to adjust taps automatically. The capacitors used for correction were rated 500kVAR, 1000kVAR and 1500kVAR, their capacitances are given by:

$$C = \frac{Q}{2\pi f V_L^2} \quad \text{--}$$

(5.1)

Where Q is the VA rating of the bank, f is the systems operating frequency in Hz and V_L is the line voltage (11kV in this case). i.e.

$$C_{500kVAR} = \frac{500'000}{2\pi(50)(11'000)^2} = 13.15uF$$

$$C_{1000kVAR} = \frac{1'000'000}{2\pi(50)(11'000)^2} = 26.31uF$$

$$C_{1500kVAR} = \frac{1'500'000}{2\pi(50)(11'000)^2} = 39.46uF$$

Table 5.4 shows the simulation results, before and after connecting the PFC capacitors.

DigSilent Power Factory Simulation Results

Operation scenario		Before PFC		After PFC	
Substation	Study Case	MVA	PF[-]	MVA	PF[-]
Bigbend	1	2.31	0.76	1.76	1.00
(Bus 24/25)					
	2	3.44	0.78	2.68	1.00
	3	4.86	0.82	3.99	1.00
	4	5.99	0.85	5.06	1.00
Maloma	1	1.40	0.49	0.72	0.96
(Bus 27)					
	2	1.71	0.70	1.22	0.98
	3	2.57	0.40	1.09	0.94

	4	3.70	0.66	2.47	0.99
Mhlume	1	0.81	0.73	0.58	1.00
(Bus 1/2)					
	2	1.63	0.74	1.21	1.00
	3	2.63	0.79	2.07	1.00
	4	3.69	0.77	2.86	0.99

Table 5.4: Simulation results of MVA and PF in substations.

NB: Though only three substations are shown in the table, but the behavior was the same in all the substations.

The table above shows a significant improvement in loading after application of PFC banks. A lower PF and higher apparent power (MVA) was experienced before correcting equipment was put into action. In some cases a unity PF was achieved while in some it couldn't be reached. This is certainly because the capacitor steps were a multiple of 500KVAR, meaning that it was impossible to compensate reactive power below 500KVAR.

For example: Maloma study case 1 where MVA=1.4, PF=0.49 implying that MW=0.69 and KVAR=1.22. Since the highest multiple of 500KVAR below 1.22MVAR is 1MVA, the 0.22MVAR won't be compensated as a leading PF is avoided thus leaving the PF below unity.

When observing the voltage profile of the 66KV busbars, we can conclude that application of static correction brought the voltage profile to acceptable levels i.e. no voltage magnitudes outside the range 0.95 to 1.05 p.u. The simulation result for maximum loading (4th case) is shown in the table below.

Bus	Before Correction		After Correction	
	kV	p.u.	kV	p.u.
21	60.44	0.916	66.97	1.015
14	67.65	1.025	68.68	1.041
26	58.89	0.892	66.56	1.009
3	62.57	0.948	65.15	0.987
12	62.70	0.950	65.17	0.987

| 18 | 63.48 | 0.962 | 65.61 | 0.994 |
| 7 | 63.06 | 0.955 | 65.38 | 0.991 |

Table 5.5: Simulation results for 66kV bus voltages.

A complete voltage profile which also include results for 11kV busbars is shown in the appendixes (Appendix C).

Line (sending - receiving)	Before Correction			After Correction		
	Vs	Vr	Current	Vs	Vr	Current
	kV	kV	kA	kV	kV	kA
14 - 21	67.65	60.44	0.09	68.68	67.13	0.06
22 - 26	60.44	58.89	0.04	66.97	66.56	0.02
14 - 17	67.65	63.48	0.17	68.68	65.61	0.14
18 - 7	63.48	63.06	0.10	65.61	65.38	0.08
7 - 4	63.06	62.57	0.03	65.38	65.15	0.02
7 - 12	63.06	62.7	0.04	65.38	65.17	0.03

Table 5.6: Simulation results for loading of 66kV transmission lines.

Note: V_r and V_s are the receiving end and sending end bus voltages, respectively.

On the 4th case we also captured simulation results on loading of the transmission line. Figure 5.8 above shows an improvement in voltage regulation due to the installation of PFC equipment. A fall in line currents was also observed.

Volt. Level [kV]	Generation [MW]/[Mvar]	Motor Load [MW]/[Mvar]	Load [MW]/[Mvar]	Compensation [MW]/[Mvar]	External Infeed [MW]/[Mvar]	Interchange to	Power Interchange [MW]/[Mvar]	Total Losses [MW]/[Mvar]	Load Losses [MW]/[Mvar]	Noload Losses [MW]/[Mvar]
11.00	0.00	0.00	22.43	0.00	0.00			0.00	0.00	0.00
	0.00	0.00	15.85	0.00	0.00			0.00	0.00	0.00
						66.00 kV	-31.26	0.04	0.04	0.00
							-23.78	2.98	2.98	0.00
						132.00 kV	8.83	0.02	0.02	0.00
							7.94	0.72	0.72	0.00
66.00	0.00	0.00	0.00	0.00	0.00			1.14	1.14	0.00
	0.00	0.00	0.00	0.00	0.00			1.83	1.83	0.00
						11.00 kV	31.23	0.04	0.04	0.00
							24.50	2.98	2.98	0.00
						132.00 kV	-32.37	0.03	0.03	0.00
							-26.33	1.55	1.55	0.00
132.00	0.00	0.00	0.00	0.00	23.59			0.00	0.00	0.00
	0.00	0.00	0.00	0.00	20.66			0.00	0.00	0.00
						11.00 kV	-8.81	0.02	0.02	0.00
							-7.22	0.72	0.72	0.00
						66.00 kV	32.40	0.03	0.03	0.00
							27.88	1.55	1.55	0.00
Total:	0.00	0.00	22.43	0.00	23.59		0.00	1.16	1.16	0.00
	0.00	0.00	15.85	0.00	20.66		0.00	4.82	4.82	0.00

Figure 5.5: Complete system report before correction.

Volt. Level [kV]	Generation [MW]/[Mvar]	Motor Load [MW]/[Mvar]	Load [MW]/[Mvar]	Compensation [MW]/[Mvar]	External Infeed [MW]/[Mvar]	Interchange to	Power Interchange [MW]/[Mvar]	Total Losses [MW]/[Mvar]	Load Losses [MW]/[Mvar]	Noload Losses [MW]/[Mvar]
11.00	0.00	0.00	22.43	-0.00	0.00			0.00	0.00	0.00
	0.00	0.00	15.85	-10.97	0.00			0.00	0.00	0.00
						66.00 kV	-30.94	0.02	0.02	0.00
							-7.88	1.81	1.81	0.00
						132.00 kV	8.51	0.01	0.01	0.00
							3.00	0.44	0.44	0.00
66.00	0.00	0.00	0.00	0.00	0.00			0.71	0.71	0.00
	0.00	0.00	0.00	0.00	0.00			1.13	1.13	0.00
						11.00 kV	30.92	0.02	0.02	0.00
							8.32	1.81	1.81	0.00
						132.00 kV	-31.64	0.02	0.02	0.00
							-9.45	0.94	0.94	0.00
132.00	0.00	0.00	0.00	0.00	23.15			0.00	0.00	0.00
	0.00	0.00	0.00	0.00	7.82			0.00	0.00	0.00
						11.00 kV	-8.50	0.01	0.01	0.00
							-2.56	0.44	0.44	0.00
						66.00 kV	31.66	0.02	0.02	0.00
							10.39	0.94	0.94	0.00
Total:	0.00	0.00	22.43	-0.00	23.15		0.00	0.72	0.72	0.00
	0.00	0.00	15.85	-10.97	7.82		0.00	2.95	2.95	0.00

Figure 5.6: Complete system report after correction.

Figure 5.5 and 5.6 shows the complete system reports before and after correction for case 4 of the simulation. The total losses of the simulation network were 1.16MW before correction in the 4th simulation case but dropped significantly to 0.72MW after the application PFC capacitors.

Also when evaluating Pprof for the 66kV part of the network based on equation 2.22 i.e. $P_{prof} = \sum_{i=1}^{N} \left(V_i - V_i^{set} \right)^2$ using the simulation results from appendix C where $V_{set} = 1pu$, we get:

Before correction: Vprof =0.032343
After correction: Vprof =0.0029550

From these calculations we can see that connection of PFC capacitors reduced Prof from 0.032343 to 0.0029550. This is advantages as we stated in the evaluation function that Pprof of zero is desirable.

5.6 Summary

In this simulation we observed a section of the network focusing mainly on five substations in the eastern side of the grid. We focused our results to four cases with each case having two operation scenarios i.e. with and without PFC equipment connected to the selected substations. The results justified the benefits of raising a low power factor correction in a power network.

CHAPTER 6: BENEFITS OF POWER FACTOR CORRECTION

6.1 Introduction

In this chapter we discuss the benefits of power factor correction to a power system. We pay particular attention to power quality and financial aspects associated with power factor correction primarily based on our simulation design and results. The payback period will also be estimated based on the five selected substations.

6.2 Improved Power Quality

The simulation results suggest that line currents will drop significantly while voltage levels get improved as a result of installation of PFC equipment. This will be very beneficial to the power system since as stated in theory (Chapter 3), high line currents are associated with large voltage drops and power losses on the network.

This will ensure that equipment operate optimally and at desired voltage levels and current, thus prolonging the life of installations and loss levels of the network are lowered during transmission. Also transformers which are probably the most valuable equipment in a power system operate at nominal tap positions during high voltage levels, thus reducing stress in coils and   insulation   damage.

6.3 Financial Optimization

It is also evident from the simulation results that after power factor correction, the amount of KVA carried by the system drops. Since the total cost of electricity involves the demand charge which is determined by the monthly maximum KVA, significant reduction in this value bring about massive savings to the power company.

Generally, the monthly cost of electricity for a large commercial or industrial customer is given by:

$$Cost = A + F + E * KWh + D * KVA_{max} \quad \text{-----------------------------------} (6.1)$$

Where:

A = Access Charge (42.92 E/Month)

F = Facility Charge (1 591.72 E/Month)

E = Energy Charge (56.86 c/KWh)

D = Demand Charge (95.46 E/KVA)

Since the access charge and energy charge are constant in every month and the real power (KW) consumed is not affected by PFC, the savings will be the difference between the apparent power (KVA_{max}) each month i.e.

$$Savings = D * (KVA_{before} - KVA_{after}) \quad \text{------------------------------------} (6.2)$$

Where:

KVA_{before} = Maximum KVA before correction

KVA_{after} = Maximum KVA after correction

The estimated savings based on 2013 loading and the suggested design for correction are tabulated on Table 6.1 below.

Substation	KVA_{before}	KVA_{after}	E(before PFC)	E(After PFC)	Savings

Bigbend	8416.70	7595.22	796,775.59	718,357.87	78,417.72
Maloma	4107.30	2952.38	385,400.98	275,151.68	110,249.30
Mhlume	7517.85	6607.45	710,971.51	624,065.36	86,906.15
Ngomane	6616.04	6013.73	624,884.72	567,388.78	57,495.94
Tabankulu	4812.56	4627.80	452,724.40	435,087.76	17,636.63

Table 6.1: Estimated monthly savings in selected substations.

The maximum KVA reading was recorded on the 11kV load bus of each of the selected substation and it is multiplied by the demand charge to determine the demand cost for each month

6.4 The Payback Period

Based on RWW Engineering LTD charges, the implementation of this project can cost approximately E8 500 000.00. This amount include:

- 5 * Nokian Power Factor Controllers **(E6 500 each)**
- 16 * 500KVAR three phase capacitor banks **(E70 000 each)**
- 1000KVAR **(E100 000)** three phase capacitor banks
- 1500 KVAR **(E125 000)** three phase capacitor banks
- Cabling **(E5 000)**
- Switch-gear and protection **(E300 000 per bank)**
- Installation and commissioning costs **(E1 500 000)**
- Transport **(E30 000)**

Summation of the monthly savings for each substations give the total monthly savings, which are

$$TotalSavings(E) = 78'417 + 110'249 + 86'906 + 57'495 + 17'636$$
$$= 350'703$$

Meaning that annually E4 208 436.00 (350 703 x 12) will be saved. With the payback given by the equation 6.3,

$$Payback\_Period = \frac{Investment}{Annual\_Savings} \text{-------------------------------- (6.3)}$$

This project will take approximately 2.02 years to recover all the cash invested in it. Since Tabankulu substation's power factor is quite good (0.96 on average), therefore it must be as well removed from the design as it brings about the least savings. This reduces the total investment o E7 700 000.00 and annual savings to E3 996 804.00, thus the payback period is reduced to 1.93 years.

6.5 Other Evaluation Methods

At SEC the cost of capital rate is 5%, therefore a positive Net Present Value (NPV) or an Internal Rate of Return (IRR) greater than 5% will mean a project is viable. The NPV and IRR over a period of 3 years are evaluated below.

- **NPV** – this method takes into account the time value of money. The estimated cash flows for each year are discounted at the discount rate and summed. For mutually exclusive projects, the one with higher NPV should be chosen for final implementation. If both give a negative answer, then they should be rejected.

  Tabankulu Substation excluded:

  $$NPV = -7'700'000 + \frac{3'996'804}{(1+0.05)^1} + \frac{3'996'804}{(1+0.05)^2} + \frac{3'996'804}{(1+0.05)^3} = 3'184'288.62$$

  Tabankulu substation included:

  $$NPV = -8'500'000 + \frac{4'208'436}{(1+0.05)^1} + \frac{4'208'436}{(1+0.05)^2} + \frac{4'208'436}{(1+0.05)^3} = 2'960'615$$

- **IRR-** this is defined as the discounted rate at which the present value of the expected future cash inflows to the present value of the project's cost are equal. Also for mutually exclusive projects, the one with higher IRR should be chosen for final implementation but if both possess an IRR that is below the stipulated cost of capital rate they should be rejected.

  Tabankulu substation excluded:

  $$-7'700'000 + \frac{3'996'804}{(1+IRR)^1} + \frac{3'996'804}{(1+IRR)^2} + \frac{3'996'804}{(1+IRR)^3} = 0 \rightarrow IRR = 25.89\%$$

  Tabankulu substation included:

  $$-8'500'000 + \frac{4'208'436}{(1+IRR)^1} + \frac{4'208'436}{(1+IRR)^2} + \frac{4'208'436}{(1+IRR)^3} = 0 \rightarrow IRR = 22.73\%$$

6.6 Summary

The installation optimize a power system technically by reducing current levels and increasing the voltage levels. It is also optimized financially by reducing operation costs through decreasing KVA requirements. On the cost and benefits analysis we can see that the payback period is around two years, but when Tabankulu substation is removed from the design the period improves. When considering the time value of money, the NPV and IRR showed promising results as the NPV was positive and the IRR was above the stipulated 5%. These two also improved on the removal of Tabankulu substation from the design.

CHAPTER 7: FINDINGS, RECOMMENDATIONS AND CONCLUSIONS

7.1 Introduction

In this chapter we check if the aims of the project were successfully fulfilled and make suggestions on what can be done to ensure that the power factor on the eastern side of the network is kept at recommended levels. We also summarize the research findings and answer the research question after which we make conclusions.

7.2 Summary of Research Findings

The power factor for four out of the five substations selected from the Eastern Grid of SEC was extremely lagging because the loads used in these industries were mainly motors, cooling equipment and irrigation pumps. These load are inductive and operate at low PF especially when they are not running at full capacity. In the year 2013 Bigbend, Maloma, Mhlume and Ngomane experienced 0.7085, 0.6569, 0.8277 and 0.8842 power factors on average respectively.

Since the primary mission of this project was to come up with a more suitable method of resolving this problem, after investigating the behavior of KVAR for each of the five substations in the year 2013 we noticed a routine in the daily loading. It came into conclusion that these four substations needed capacitors for correction since the load was not too dynamic, so four bank steps in each could raise the PF to 0.98 lagging on average. This correction will optimize the system both technically and financially through a reduced magnitude of line currents, raised voltage levels and a reduction in KVA consumption.

When doing a cost benefit analysis based on the cost of the correction equipment and savings which will be obtained after implementation of the project it came into attention that Tabankulu need to be excluded from the design. This is because its PF was 0.97 on average before correction, so installation of PFC equipment brought lesser improvements and small saving. As a result it prolonged the payback period and reduced IRR and NVP.

7.3 Recommendations

Informed by the theories and findings observed while doing this research, we will like to make the following recommendations:

i.   It is important that capacitor banks used in correction have enough electrical steps to meet the demands of PF. On the other hand, too many steps can be problematic as switching causes transients. The IEC 831 standard also state that the maximum number of switching activities should not exceed 5000 per annum. Therefore a proper decision on the number of steps to be used should be determined through a comprehensive study about the load behavior. The use of fast switching contactors with low de-bounce having pre-charging resistors to damp the inrush current could be the solution.

ii. The PFC system should also have a maintenance plan. This should involve interrogation of the equipment before replacements of malfunctioning components, removal of foreign matter etc. can be done to ensure optimal operation and that banks do not fail until their expected lifespan.

iii. Swaziland is still developing and the population is always increasing, so there is a need on continuous monitoring of the load. This will ensure that additional PFC equipment is introduced immediately when consumption is increased or when a new industry is introduced.

iv. Non-industrial customers should be also advised to implement PFC in their homesteads. Residential loads like fluorescent lamps, refrigerators, fridges etc. also operate at low PF, so practicing correction residentially can further enhance the reliability and quality of the power system.

7.4 Conclusion

Power Factor Correction is a vital requirement especially for industrial and commercial loads operating at a low PF. This ensures that a greater percentage KVAR in transmission/ distribution lines is offset, thus increasing the network's capacity. In order to figure out where to implement correction, an in-depth study of the load behavior is a necessity.

The main objectives of this research were to identify the causes of a low PF in SEC's network, calculate the PF and component values for PFC equipment, evaluate the impact of a low PF and identify the current practice at SEC in as far as correction is done. In fulfilling these goals, five substations were picked from the eastern grid. A model for the solution have been proposed after this research.

The DigSilent simulation of power factor correction was done successfully. The simulation model was designed to be as close as possible to the real network. Each component of the model had its type properly configured to its real time specifications before the simulation was done. The developed model supported numerous operational scenarios and easy exportation of simulation results for analysis.

The simulation results showed that the power system could operate more efficiently if this proposed model is put into action. When the correction equipment were enabled on the design, a significant drop in line currents and KVA requirement of the network were observed. On the other hand bus voltage levels increased. This shows that PFC optimizes the power system as power losses and voltage drops were greatly reduced.

The research hypothesis was to prove or otherwise that if we know the factors that contribute to the power factor being low and the load behavior of the affected portion of the network, then we can choose the appropriate PFC method to minimize the effects of those loads. Although we worked on a small portion of the network, we can indeed conclude that a comprehensive study of a load profile is key towards choosing the correct PFC method for any substation.

REFERENCES

[1] T. L. Skvarenina and W. E. Dewitt, Electrical Power and Control, 2004, pp. 75-90.

[2] S. A. Nasar and F. C. Trutt, Power Systems, 1998, p. 21.

[3] E. Hughes, "Power in AC Circuits," in *Electrical and Electronic Technology*, 2008, pp. 259-271.

[4] J. D. Clover, M. S. Sarma and T. J. Overbye, Power System Analysis and Design, Cengage Learning, 2012, p. 55.

[5] A. Bhatia, "PHDonline course," 2012. [Online]. Available: www.PDHonline.org/courses/e144/e144content.pdf. [Accessed 12 September 2013].

[6] N. G. Hingorani and L. Gyugyi, "IEEE Press book," *Understanding Facts: Concepts and Technology of Flexible AC Transmission Systems*, 2000.

[7] M. T. Yada, J. A. Zanzarukya, N. H. Patel, S. H. Domaliya and A. J. Lathigaya, "Static Var Compensator," LE college, Morbi, 2013.

[8] A. M. Kamarposhti and M. Alinezhad, "Comparison of SVC and STATCOM in Static Voltage Stability Margin Enhancement," *International Journal of Electrical and Electronics Engineering*, vol. 4, no. 5, pp. 324-325, 2010.

[9] Matlab, "Static Var Compensator (Phasor Type)".

[10] "Power Factor Forrection," [Online]. Available: http://www.nhp.com.au/files/editor_upload/File/Power%20Quality/Introduction-to-Power-Factor-Correction.pdf. [Accessed 1 October 2013].

[11] "Sustainable Power Systems," [Online]. Available: http://www.sustainablepowersystems.com/wp-content/uploads/2012/09/Synchronous-Condenser-Rev-3.pdf. [Accessed 13 October 2013].

[12] M. Pikulski, "CONTROLLED SOURCES OF REACTIVE POWER," Aalborg University, Aalborg, 2008.

[13] "ABB STATCOM For flexibility in power systems," [Online]. Available: http://www05.abb.com/global/scot/scot256.nsf/veritydisplay/26ab4cd0ecbe3bcbc1256b9d004a7c88/$file/statcom.pdf. [Accessed 14 October 2013].

[14] Matlab, "STATCOM Power Component".

[15] H. Seifi and M. S. Sepasian, Electric Power System Planning, Tehran: Springer, 2011.

[16] Siemens, "Siemens Global Website," Siemens, 2002. [Online]. Available: http://www.energy.siemens.com/hq/en/power-transmission/facts/static-var-compensator-classic/. [Accessed 23 September 2013].

[17] T. R. Kuphaldt, *Lessons In Electric Circuits, Volume II – AC,* 2007.

APPENDIXES

Appendix A: Load Profiles

Load Beahaviour of Maloma Substation

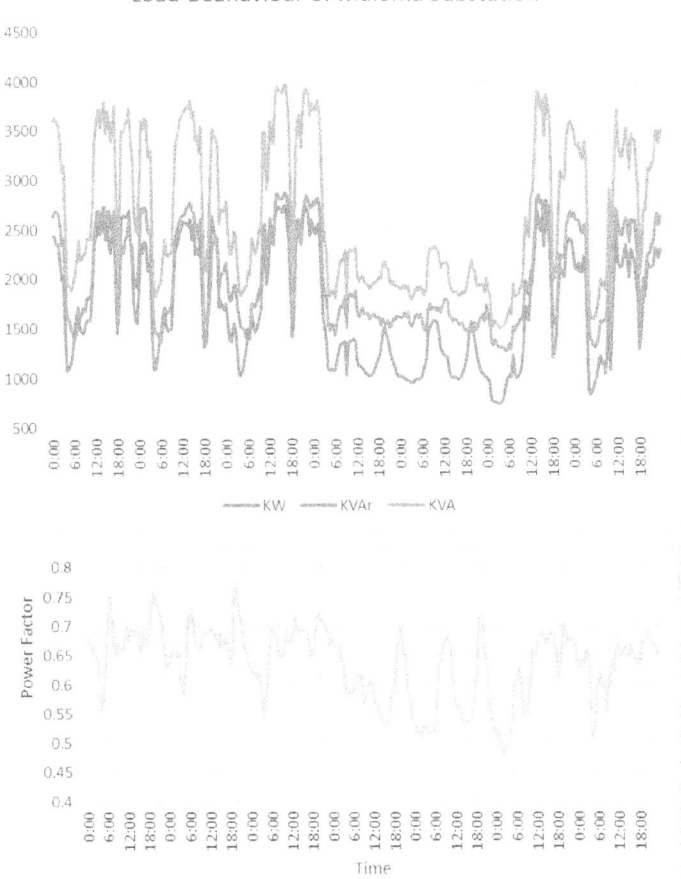

Figure 4.1: Load profile for Maloma substation (14-20Nov 2012).

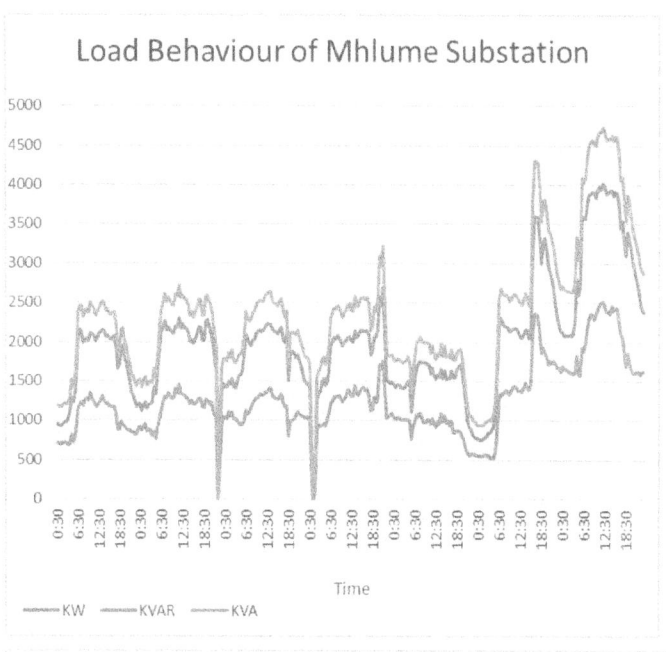

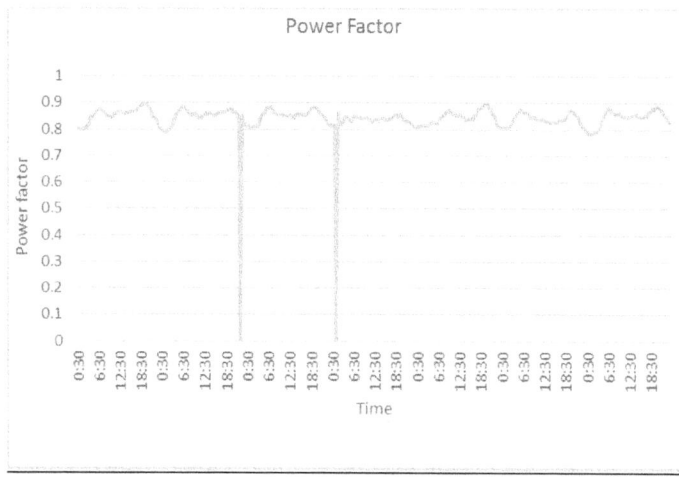

Figure 4.2: Load profile for Mhlume substation (14-20Nov 2012).

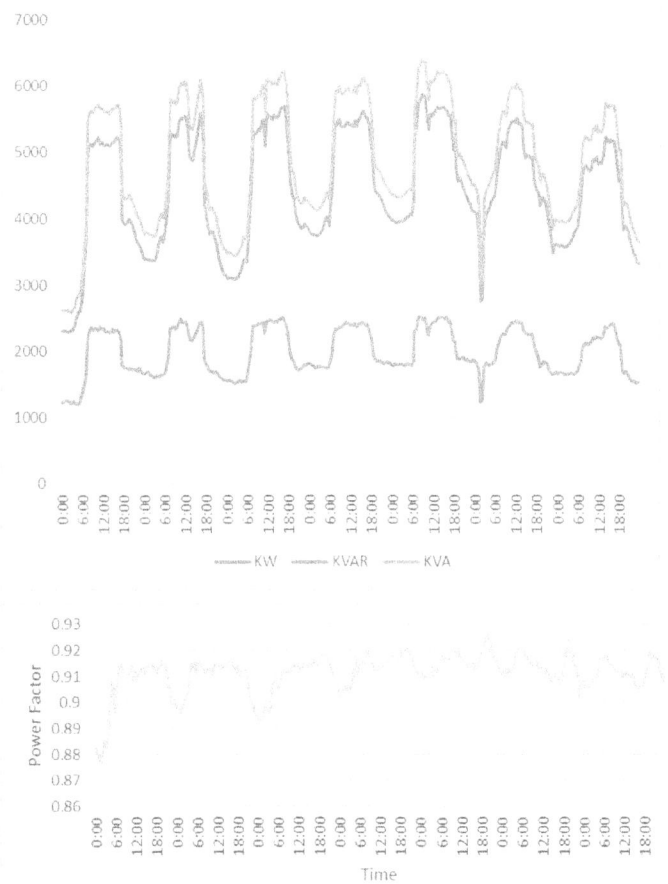

Figure 4.3: Load profile for Ngomane substation (14-20Nov 2012).

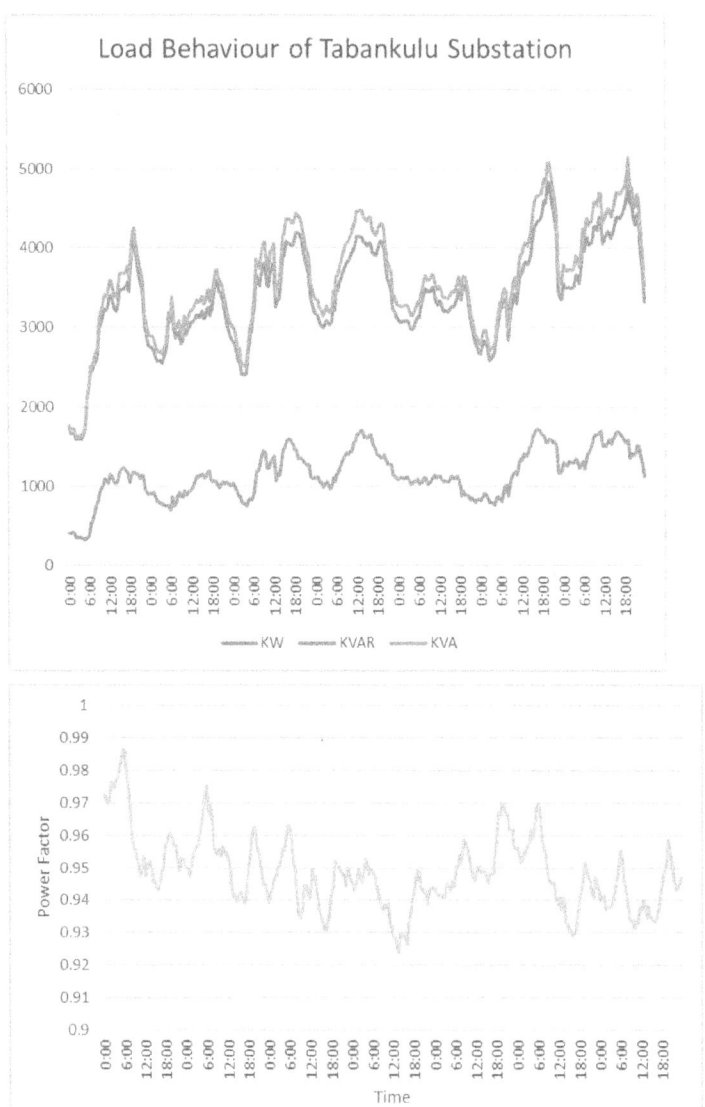

Figure 4.4: Load profile for Tabankulu substation (14-20Nov 2012).

Appendix B – Substations

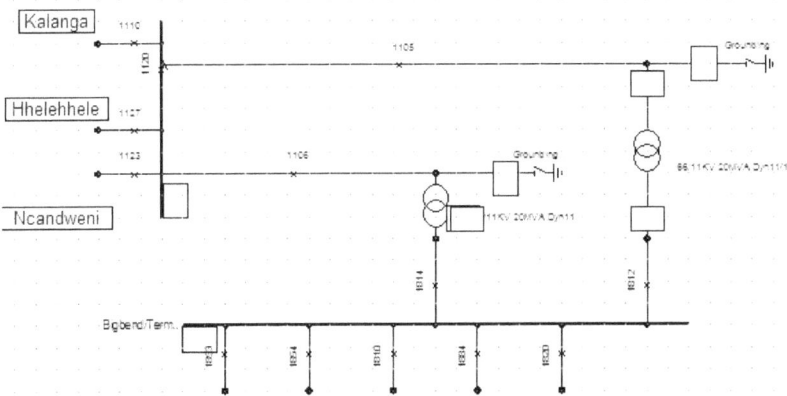

Figure 1: Bigbend Substation

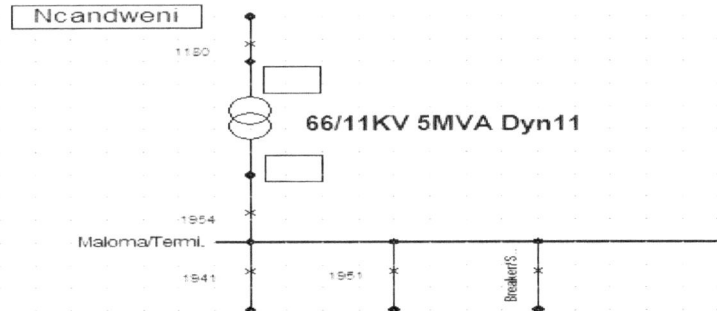

Figure 2: Maloma Substation

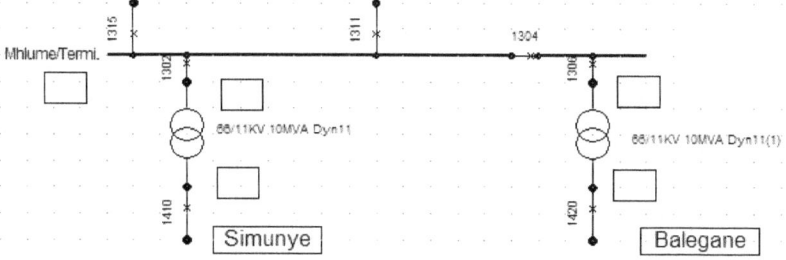

Figure 3: Mhlume Substation

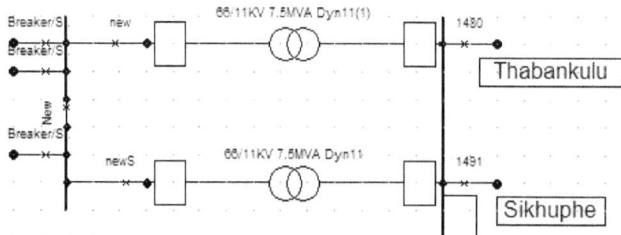

Figure 4: Ngomane Substation

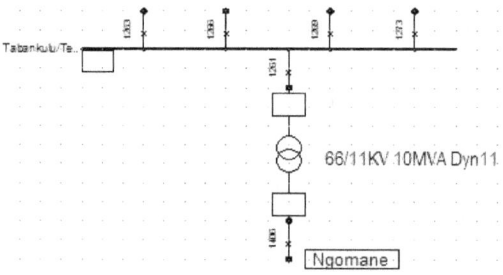

Figure 5: Tabankulu Substation

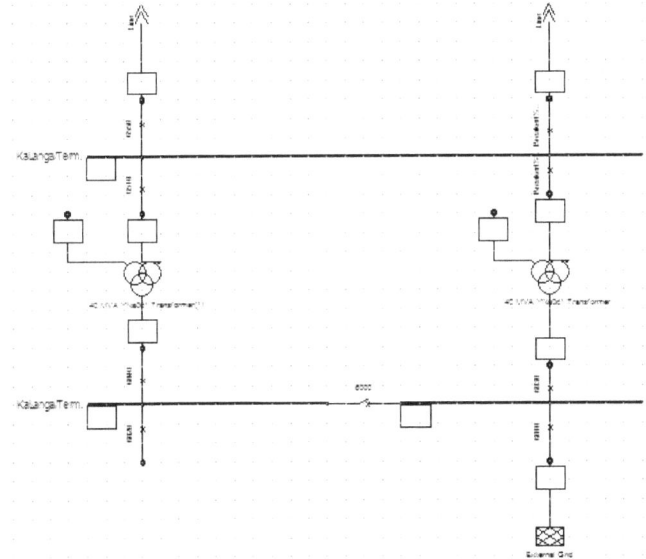

Figure 6: Kalanga Substation.

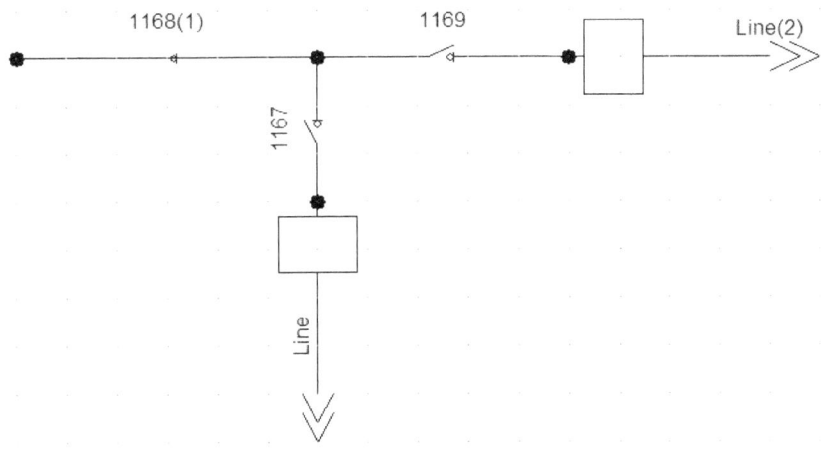

Figure 7: Ncandweni Switchyard.

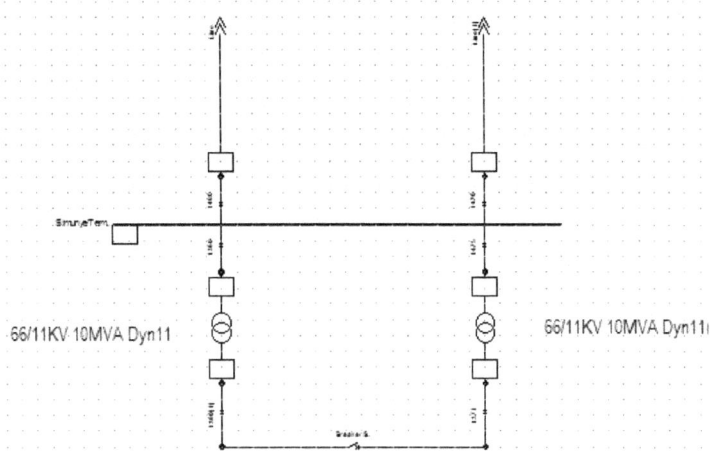

Figure 8: Simunye Substation.

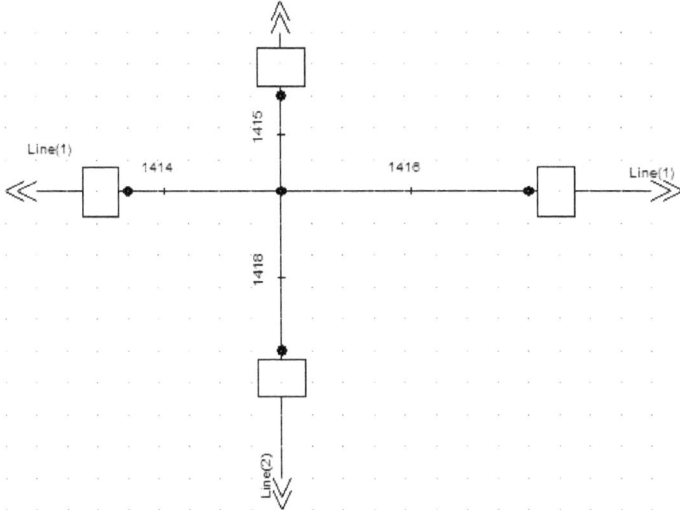

Figure 9: Tabankulu Switchyard.

Appendix C : Simulation Results

- ## Case 4: Capacitor banks disconnected

```
---------------------------------------------------------------------------------------
|           |                              |  DIgSILENT  | Project:                      |
|           |                              |  PowerFactory |------------------------------
|           |                              |  15.0.2     | Date:  5/1/2014               |
---------------------------------------------------------------------------------------

---------------------------------------------------------------------------------------
| Load Flow Calculation              Complete System Report: Substations, Voltage Profiles, Grid Interchange |
---------------------------------------------------------------------------------------

|   AC Load Flow, balanced, positive sequence      |  Automatic Model Adaptation for Convergence      No   |
|   Automatic Tap Adjust of Transformers     Yes   |  Max. Acceptable Load Flow Error for                  |
|   Consider Reactive Power Limits           Yes   |    Nodes                                      1.00 kVA |
|                                                  |    Model Equations                            0.10 %  |
---------------------------------------------------------------------------------------

---------------------------------------------------------------------------------------
| Grid: Eatsern Grid 1    System Stage: Eatsern Grid 1  | Study Case: Case 4          | Annex:         / 4 |
---------------------------------------------------------------------------------------

|            rtd.V    Bus - voltage                          Voltage - Deviation [%]                     |
|            [kV] [p.u.]   [kV] [deg]        -10      -5           0          +5        +10              |
---------------------------------------------------------------------------------------

|Bigbend                                                                                                 |
| Bus 21     66.00  0.916  60.44  -5.65            <<<<<<<<<<<<<<<<<<<<|                                  |
| Bus 23     11.00  0.933  10.27  -4.81            <<<<<<<<<<<<<<<<|                                      |
| Bus 22     66.00  0.969  63.95  -2.39                     <<<<<<<|                                      |
|KaLanga                                                                                                 |
| Bus 15    132.00  1.050 138.60   0.00                             |>>>>>>>>>>>>                         |
| Bus 16    132.00  1.050 138.60   0.00                             |>>>>>>>>>>>>                         |
| Bus 14     66.00  1.025  67.65  -1.58                             |>>>>>>                               |
|Maloma                                                                                                  |
| Bus 27     11.00  0.843   9.27  -8.93        <\\\\\\\\\\\\\\\\\\\\\\\\\|                                |
| Bus 26     66.00  0.892  58.89  -6.21        <\\\\\\\\\\\\\\\\\\\\\\\|                                  |
|Mhlume                                                                                                  |
| Bus 1      11.00  0.924  10.17  -4.97            <<<<<<<<<<<<<<<<<<|                                    |
| Bus 2      11.00  0.924  10.17  -4.97            <<<<<<<<<<<<<<<<<<|                                    |
```

Bus 3	66.00	0.948	62.57	-3.28	<<<<<<<<<<<<<	
Bus 4	66.00	0.924	61.01	-4.97	<<<<<<<<<<<<<<<<<<	
Ngomane						
Bus 12	66.00	0.950	62.70	-3.28	<<<<<<<<<<<<	
Bus 9	11.00	0.938	10.32	-4.66	<<<<<<<<<<<<<<<	
Bus 10	11.00	0.938	10.32	-4.66	<<<<<<<<<<<<<<<	
Simunye						
Bus 18	66.00	0.962	63.48	-2.94	<<<<<<<<<	
Bus 19	11.00	0.962	10.58	-2.94	<<<<<<<<<	
Bus 20	11.00	0.918	10.10	-6.39	<<<<<<<<<<<<<<<<<<<<	
Tabankulu						
Bus 6	11.00	0.898	9.88	-4.37	<<<<<<<<<<<<<<<<<<<<<<<<<<	
Bus 7	66.00	0.955	63.06	-3.10	<<<<<<<<<<<<	

| | | | DIgSILENT | Project: |
| | | | PowerFactory |----------------------
| | | | 15.0.2 | Date: 5/1/2014 |

| Load Flow Calculation   Complete System Report: Substations, Voltage Profiles, Grid Interchange |

AC Load Flow, balanced, positive sequence	Automatic Model Adaptation for Convergence No
Automatic Tap Adjust of Transformers Yes	Max. Acceptable Load Flow Error for
Consider Reactive Power Limits Yes	Nodes 1.00 kVA
	Model Equations 0.10 %

| Grid: Eatsern Grid 1   System Stage: Eatsern Grid 1 | Study Case: Case 4 | Annex: / 5 |

Volt.	Generation	Motor	Load	Compen-	External		Power	Total	Load	Noload	
Level		Load		sation	Infeed	Interchange	Interchange	Losses	Losses	Losses	
	[MW]/	[MW]/	[MW]/	[MW]/	[MW]/	to	[MW]/	[MW]/	[MW]/	[MW]/	
[kV]	[Mvar]	[Mvar]	[Mvar]	[Mvar]	[Mvar]		[Mvar]	[Mvar]	[Mvar]	[Mvar]	

11.00	0.00	0.00	22.43	0.00	0.00			0.00	0.00	0.00	
	0.00	0.00	15.85	0.00	0.00			0.00	0.00	0.00	
						66.00 kV	-31.26	0.04	0.04	0.00	
							-23.78	2.98	2.98	0.00	
						132.00 kV	8.83	0.02	0.02	0.00	

							7.94	0.72	0.72	0.00	
66.00	0.00	0.00	0.00	0.00	0.00			1.14	1.14	0.00	
	0.00	0.00	0.00	0.00	0.00			1.83	1.83	0.00	
						11.00 kV	31.23	0.04	0.04	0.00	
							24.50	2.98	2.98	0.00	
						132.00 kV	-32.37	0.03	0.03	0.00	
							-26.33	1.55	1.55	0.00	
132.00	0.00	0.00	0.00	0.00	23.59			0.00	0.00	0.00	
	0.00	0.00	0.00	0.00	20.66			0.00	0.00	0.00	
						11.00 kV	-8.81	0.02	0.02	0.00	
							-7.22	0.72	0.72	0.00	
						66.00 kV	32.40	0.03	0.03	0.00	
							27.88	1.55	1.55	0.00	
Total:	0.00	0.00	22.43	0.00	23.59		0.00	1.16	1.16	0.00	
	0.00	0.00	15.85	0.00	20.66		0.00	4.82	4.82	0.00	

			DIgSILENT	Project:	
			PowerFactory		------------------------------
			15.0.2	Date: 5/1/2014	

Load Flow Calculation	Complete System Report: Substations, Voltage Profiles, Grid Interchange

AC Load Flow, balanced, positive sequence	Automatic Model Adaptation for Convergence	No
Automatic Tap Adjust of Transformers	Yes	Max. Acceptable Load Flow Error for	
Consider Reactive Power Limits	Yes	Nodes	1.00 kVA
		Model Equations	0.10 %

Total System Summary	Study Case: Case 4	Annex:	/ 6

Generation	Motor Load	Load	Compen- sation	External Infeed		Inter Area Flow	Total Losses	Load Losses	Noload Losses	
[MW]/ [Mvar]	[MW]/ [Mvar]	[MW]/ [Mvar]	[MW]/ [Mvar]	[MW]/ [Mvar]		[MW]/ [Mvar]	[MW]/ [Mvar]	[MW]/ [Mvar]	[MW]/ [Mvar]	

| \Muzi\PFCdemo(1)\Network Model\Network Data\Eatsern Grid 1 | |

| | 0.00 | 0.00 | 22.43 | 0.00 | 23.59 | 0.00 | 1.16 | 1.16 | 0.00 | |
| | 0.00 | 0.00 | 15.85 | 0.00 | 20.66 | 0.00 | 4.82 | 4.82 | 0.00 | |

| Total: | |

| | 0.00 | 0.00 | 22.43 | 0.00 | 23.59 | 1.16 | 1.16 | 0.00 | |
| | 0.00 | 0.00 | 15.85 | 0.00 | 20.66 | 4.82 | 4.82 | 0.00 | |

- **Case 4: Capacitor banks active**

```
-------------------------------------------------------------------------------------
|                 |                                        |   DIgSILENT   | Project:
|
|                 |                              | PowerFactory  |-------------------------------
-
|                 |                                        |   15.0.2   | Date:  5/1/2014
|
-------------------------------------------------------------------------------------
-------------------------------------------------------------------------------------
| Load Flow Calculation                 Complete System Report: Substations, Voltage Profiles, Grid Interchange
|
-------------------------------------------------------------------------------------
|    AC Load Flow, balanced, positive sequence       |   Automatic Model Adaptation for Convergence         No
|
|       Automatic Tap Adjust of Transformers            Yes    |    Max. Acceptable Load Flow Error for
|
|    Consider Reactive Power Limits            Yes   |   Nodes                                        1.00 kVA
|
|                                              |    Model Equations                               0.10 %
|
-------------------------------------------------------------------------------------
-------------------------------------------------------------------------------------
| Grid: Eatsern Grid 1     System Stage: Eatsern Grid 1  | Study Case: Case 4         | Annex:            / 4
|
-------------------------------------------------------------------------------------
|                  rtd.V     Bus - voltage                            Voltage - Deviation [%]
|
|                 [kV] [p.u.]    [kV]  [deg]           -10      -5       0      +5      +10
|
-------------------------------------------------------------------------------------
|Bigbend
|
|   Bus 21              66.00   1.015    66.97  -6.17                                      |>>>>
|
```

```
|   Bus 23              11.00   1.017   11.18   -5.48                      |>>>>
|
|   Bus 22              66.00   1.017   67.13   -3.37                      |>>>>
|
|KaLanga
|
|   Bus 15             132.00   1.050  138.60    0.00             |>>>>>>>>>>>>
|
|   Bus 16             132.00   1.050  138.60    0.00             |>>>>>>>>>>>>
|
|   Bus 14              66.00   1.041   68.68   -1.54             |>>>>>>>>>>>
|
|Maloma
|
|   Bus 27              11.00   1.004   11.04   -8.90                        |>
|
|   Bus 26              66.00   1.009   66.56   -6.88                       |>>
|
|Mhlume
|
|   Bus  1              11.00   0.983   10.81   -5.40                    <<<<|
|
|   Bus  2              11.00   0.983   10.81   -5.40                    <<<<|
|
|   Bus  3              66.00   0.987   65.15   -3.87                     <<<|
|
|   Bus  4              66.00   0.983   64.86   -5.40                    <<<<|
|
|Ngomane
|
|   Bus 12              66.00   0.987   65.17   -3.85                     <<<|
|
|   Bus  9              11.00   0.986   10.85   -5.10                     <<<|
|
|   Bus 10              11.00   0.986   10.85   -5.10                     <<<|
|
```

| Simunye
|
| Bus 18   66.00   0.994   65.61   -3.36   <|
|
| Bus 19   11.00   0.994   10.94   -3.36   <|
|
| Bus 20   11.00   0.952   10.48   -6.58   <<<<<<<<<<<|
|
| Tabankulu
|
| Bus 6   11.00   0.976   10.74   -4.69   <<<<<<|
|
| Bus 7   66.00   0.991   65.38   -3.60   <<|
|

| | | DIgSILENT | Project:
|
| | | PowerFactory |---------------------------------
-
| | | 15.0.2 | Date: 5/1/2014
|

| Load Flow Calculation   Complete System Report: Substations, Voltage Profiles, Grid Interchange
|

| AC Load Flow, balanced, positive sequence | Automatic Model Adaptation for Convergence   No
|
| Automatic Tap Adjust of Transformers   Yes | Max. Acceptable Load Flow Error for
|
| Consider Reactive Power Limits   Yes | Nodes   1.00 kVA
|
| | Model Equations   0.10 %
|

Volt.	Generation	Motor	Load	Compen-	External		Power	Total	Load	Noload
Level		Load		sation	Infeed	Interchange	Interchange	Losses	Losses	Losses
	[MW]/	[MW]/	[MW]/	[MW]/	[MW]/	to	[MW]/	[MW]/	[MW]/	[MW]/
[kV]	[Mvar]	[Mvar]	[Mvar]	[Mvar]	[Mvar]		[Mvar]	[Mvar]	[Mvar]	[Mvar]
11.00	0.00	0.00	22.43	-0.00	0.00			0.00	0.00	0.00
	0.00	0.00	15.85	-10.97	0.00			0.00	0.00	0.00
						66.00 kV	-30.94	0.02	0.02	0.00
							-7.88	1.81	1.81	0.00
						132.00 kV	8.51	0.01	0.01	0.00
							3.00	0.44	0.44	0.00
66.00	0.00	0.00	0.00	0.00	0.00			0.71	0.71	0.00
	0.00	0.00	0.00	0.00	0.00			1.13	1.13	0.00
						11.00 kV	30.92	0.02	0.02	0.00
							8.32	1.81	1.81	0.00
						132.00 kV	-31.64	0.02	0.02	0.00
							-9.45	0.94	0.94	0.00

```
---------------------------------------------------------------------------------------------------------
| 132.00    0.00    0.00    0.00    0.00    23.15                          0.00    0.00    0.00
|
|          0.00    0.00    0.00    0.00     7.82                          0.00    0.00    0.00
|
|                                                  11.00 kV    -8.50    0.01    0.01    0.00
|
|                                                              -2.56    0.44    0.44    0.00
|
|                                                  66.00 kV    31.66    0.02    0.02    0.00
|
|                                                              10.39    0.94    0.94    0.00
|
---------------------------------------------------------------------------------------------------------
| Total:    0.00    0.00   22.43   -0.00    23.15               0.00    0.72    0.72    0.00
|
|          0.00    0.00   15.85  -10.97     7.82               0.00    2.95    2.95    0.00
|
---------------------------------------------------------------------------------------------------------
---------------------------------------------------------------------------------------------------------
|             |                                              |   DIgSILENT   | Project:
|
|          |                                       | PowerFactory  |-------------------------------
-
|          |                                       |   15.0.2   | Date:  5/1/2014
|
---------------------------------------------------------------------------------------------------------
---------------------------------------------------------------------------------------------------------
| Load Flow Calculation                    Complete System Report: Substations, Voltage Profiles, Grid Interchange
|
---------------------------------------------------------------------------------------------------------
|   AC Load Flow, balanced, positive sequence     |   Automatic Model Adaptation for Convergence        No
|
|      Automatic  Tap  Adjust  of  Transformers          Yes    |     Max.  Acceptable  Load  Flow  Error  for
|
|   Consider Reactive Power Limits           Yes   |    Nodes                          1.00 kVA
|
```

| Total System Summary | Study Case: Case 4 | Annex: | / 6

Generation	Motor	Load	Compen-	External		Inter Area	Total	Load	Noload
		Load		sation	Infeed	Flow	Losses	Losses	Losses
[MW]/	[MW]/	[MW]/	[MW]/	[MW]/		[MW]/	[MW]/	[MW]/	[MW]/
[Mvar]	[Mvar]	[Mvar]	[Mvar]	[Mvar]		[Mvar]	[Mvar]	[Mvar]	[Mvar]

\Muzi\PFCdemo(1)\Network   Model\Network   Data\Eatsern   Grid   1

| 0.00 | 0.00 | 22.43 | -0.00 | 23.15 | | 0.00 | 0.72 | 0.72 | 0.00 |
| 0.00 | 0.00 | 15.85 | -10.97 | 7.82 | | 0.00 | 2.95 | 2.95 | 0.00 |

Total:

| 0.00 | 0.00 | 22.43 | -0.00 | 23.15 | | | 0.72 | 0.72 | 0.00 |
| 0.00 | 0.00 | 15.85 | -10.97 | 7.82 | | | 2.95 | 2.95 | 0.00 |

www.ingramcontent.com/pod-product-compliance
Lightning Source LLC
Chambersburg PA
CBHW071207220526
45468CB00002B/527